Plant Propagation and Nursery Management for Fruit Crops

NIPA® GENX ELECTRONIC RESOURCES & SOLUTIONS P. LTD.

New Delhi-110 034

About the Editors

Dr. Rakesh Kumar Jat is currently working as Assistant Professor, Department of Fruit Science, College of Horticulture, Sardarkrushinagar Dantiwada Agricultural University, Jagudan, Mehsana, Gujarat. He has completed his B.Sc. (Agriculture) Honours from RAU, Bikaner (Rajasthan), M.Sc. Fruit Science (Hort.) from NAU, Navsari (Gujarat) and completed Ph.D. in Fruit Science (Hort.) from Sardarkrushinagar Dantiwada Agricultural University, Jagudan, Mehsana, (Gujarat). He has more than 9 years of experience in teaching, research and extension. Dr. Jat is currently guiding 5 M.Sc. students and handling several research projects. He has been awarded Young Scientist Award (2022). He has published several research papers in reputed national and international journals, book chapters and popular articles and also contributed to several recommendations for farming community and nurserymen.

Mohan Lal Jat is pursuing Ph.D in Department of Horticulture, Chaudhary Charan Singh Haryana Agricultural University, Hisar, Haryana. He has completed his B.Sc. (Horticulture) Honours from Agriculture University, Kota (Rajasthan), M.Sc. Ag. (Horticulture) from Govind Ballabh Pant University of Agriculture and Technology, Pantnagar, Udham Singh Nagar (Uttarakhand). He qualified JRF (21st rank), SRF (2nd rank) and NET exam conducted by ICAR. He has published 10 research papers in reputed national and international journals, 4 book chapters and 15 popular articles. He is also a lifetime member of several scientific societies.

Plant Propagation and Nursery Management for Fruit Crops

Rakesh Kumar Jat
Assistant Professor, Department of Fruit Scienc
College of Horticulture
Sardarkrushinagar Dantiwada Agricultural University
Jagudan, Mehsana, Gujarat

Mohan Lal Jat
Research Scholar, Department of Horticulture
College of Agriculture
Chaudhary Charan Singh Haryana Agricultural University
Hisar, Haryana

NIPA® GENX ELECTRONIC RESOURCES & SOLUTIONS P. LTD.
New Delhi-110 034

NIPA® GENX ELECTRONIC RESOURCES & SOLUTIONS P. LTD.

101,103, Vikas Surya Plaza, CU Block
L.S.C. Market, Pitam Pura, New Delhi-110 034
Ph : +91 11 27341616, 27341717, 27341718
E-mail: newindiapublishingagency@gmail.com
www: www.nipabooks.com

For customer assistance, please contact
Phone: + 91-11-27 34 17 17
Fax: + 91-11-27 34 16 16
E-Mail: feedbacks@nipabooks.com

ISBN: 978-93-95319-64-5

Composed and Designed by NIPA®.

Preface

Agricultural and horticultural sciences are essential to our existence, since they provide the most basic need of food. Generally, qualitative and quantitative food can only be produced from healthy plants, which can only be started from healthy seedlings or saplings. A diverse edapho-climatic condition offers the opportunity to plant different species in India. A number of orchard trees have typically been planted in kitchen gardens and on fields, which is a well-established method for obtaining quality planting materials. Consequently, the requirement for high quality planting materials in India varies and also is perpetual because of the varied edapho-climatic conditions. However, the species or varieties or genotypes that are suitable for cultivation in one region may not be remunerative in another. It is thus possible to increase agriculture productivity by developing site specific quality planting materials. The availability of quality planting materials at lower prices is conducive to large scale plantings. It is possible to establish sustainable agriculture by focusing on the production of quality planting materials in this stage. Nursery management can be an important tool for successfully executing the activity in order to produce quality planting material with minimum input. Additionally, providing nutritional security for an expanding population is another challenge for agrarian communities. Use of qualitative nursery planting materials will lead to increased production of horticultural crops, thus solving the challenging task.

Thus, the efforts have been made so to compile the book entitled "Plant Propagation and Nursery Management for Fruit Crops" for students, teachers, research scholars as well as budding horticulturists. This book covers most of the areas of plant propagation including from traditional to recent advances. The emphasis has not only been laid on plant propagation techniques and nursery management practices, but also on the ideas and their applications. Special care has been taken so that the rigor of science is not lost while simplifying the language. We hope and trust that this book will help the students to understand various techniques of plant propagation and nursery management and their utility in a very simple way. An attempt has been made to design chapters of the book based on the current curricula and syllabi of ICAR.

We are thankful to our respective home institutes; College of Horticulture, Sardarkrushinagar Dantiwada Agricultural University, Jagudan, Mehsana, Gujarat and Chaudhary Charan Singh Haryana Agricultural University, Hisar, Haryana for invariable motivation in the preparation of this text. The constant encouragement and constructive criticism have motivated us to bring this manuscript to the present form. We are very grateful for a number of friends and colleagues in encouraging us to start the work, persevere with it and finally to publish the book. Devotion and dedication of esteemed authors of different chapters of the book is highly acknowledged for preparing, reviewing, refining and finalization of the manuscript in a set pattern and helping to bring it in the presentable form. We are highly thankful to the publisher for accepting our proposal and publishing the book without any complexities and delay.

Last but not the least; we owe the obvious support and sacrifice of our families. It would be unrealistic to assume that the book is free from errors and the suggestions will be highly appreciated for making improvement in this manuscript.

Editors

Contents

List of Contributors

Ajit Kumar Singh, Division of Fruits and Horticultural Technology, Indian Agricultural Research Institute, New Delhi

Ashok Dhakad, Division of Fruits and Horticultural Technology, Indian Agricultural Research Institute, New Delhi

G.S. Patel, Department of Vegetable Science, College of Horticulture, Sardarkrushinagar Dantiwada Agricultural University, Jagudan, Mehsana, Gujarat

Jitendra Singh Shivran, Department of Horticulture, College of Agriculture, G.B. Pant University of Agriculture and Technology, Pantnagar, Udham Singh Nagar, Uttarakhand

Kapil Mohan Sharma, Date palm Research Station, Sardarkrushinagar Dantiwada Agricultural University, Mundra-Kachchh, Gujarat

M.K. Sharma, Department of Vegetable Science, College of Horticulture, Sardarkrushinagar Dantiwada Agricultural University, Jagudan, Mehsana, Gujarat

M.L. Jat, Department of Horticulture, College of Agriculture, Chaudhary Charan Singh Haryana Agricultural University, Hisar, Haryana

Mukesh Chand Bhateswar, Department of Horticulture, SKN College, Sri Karan Narendra Agriculture University, Jobner, Jaipur, Rajasthan

Mukesh Kumar, Department of Natural Resource Management, College of Horticulture, Sardarkrushinagar Dantiwada Agricultural University, Jagudan, Mehsana, Gujarat

O.P. Kumawat, Department of Horticulture, Rajasthan College of Horticulture, Maharana Pratap University of Agriculture and Technology, Udaipur, Rajasthan

Pankaj Yadav, Department of Pathology, College of Agriculture, Chaudhary Charan Singh Haryana Agricultural University, Hisar, Haryana

Pooja Sharma, Department of Horticulture, SKN College, Sri Karan Narendra Agriculture University, Jobner, Jaipur, Rajasthan

Poornita Raturi, Department of Horticulture, College of Agriculture, G. B. Pant University of Agriculture and Technology, Pantnagar, Udham Singh Nagar, Uttarakhand

R.K. Jat, Department of Fruit Science, College of Horticulture, Sardarkrushinagar Dantiwada Agricultural University, Jagudan, Mehsana, Gujarat

Rajender Kumar, Department of Horticulture, College of Agriculture, G. B. Pant University of Agriculture and Technology, Pantnagar, Udham Singh Nagar, Uttarakhand

Rajesh Mor, Department of Fruit Science, College of Horticulture, Maharana Pratap Horticultural University, Karnal, Haryana

Rajkumar Jat, Department of Horticulture, College of Agriculture, G. B. Pant University of Agriculture and Technology, Pantnagar, Udham Singh Nagar, Uttarakhand

Ravi Kumar, Department of Horticulture, College of Agriculture, G. B. Pant University of Agriculture and Technology, Pantnagar, Udham Singh Nagar, Uttarakhand

S.K. Acharya, Department of Vegetable Science, College of Horticulture, Sardarkrushinagar Dantiwada Agricultural University, Jagudan, Mehsana, Gujarat

Sampurna Nand Singh, Department of Horticulture, College of Agriculture, G. B. Pant University of Agriculture and Technology, Pantnagar, Udham Singh Nagar, Uttarakhand

Shreya, Department of Basic Science, College of Horticulture, Sardarkrushinagar Dantiwada Agricultural University, Jagudan, Mehsana, Gujarat

Shubham Jagga, Division of Fruits and Horticultural Technology, Indian Agricultural Research Institute, New Delhi

Sonu Kumar, Department of Horticulture, College of Agriculture, Chaudhary Charan Singh Haryana Agricultural University, Hisar, Haryana

Vivek Saurabh, Division of Fruits and Horticultural Technology, Indian Agricultural Research Institute, New Delhi

Terminology

Adventitious organs: Organs that rise from the dedifferentiation of parenchyma cells; when they originate from callus (also composed of parenchyma cells) their organogenesis is termed indirect.

Adventitious roots: Roots that arise on aerial plant parts, underground stems and old root parts.

Adventitious buds (and shoots): Arise from any plant part other than terminal, lateral, or latent buds on stems.

Anaphase: The mitotic spindle fiber microtubules attached to each chromosome pair at the centromere contract, pulling the chromosomes apart.

Amphimicts: A term used to indicate the fact that both parents contribute to the inheritance of their offspring.

Annuals: Plants that perform their entire life cycle from seed to flower to seed within a single growing season.

Apomixis: It is a phenomenon in which there is development of seed without fertilization or sexual fusion.

Approach graftage: The root system of the scion and shoot system of the rootstock are not removed until after successful graft union formation occurs.

Bench budding (bench working): Budding under protected culture onto a dormant rootstock using a bench or onto a containerized rootstock that may be active or dormant.

Budding: A form of grafting in which only one bud is inserted in the rootstock.

Callus: A mass of undifferentiated parenchymatous cells.

Chimera: A condition in which an individual contains different genotypes or distinct genotypes growing side-by-side within the same plant.

Chromosome: Structures within the nucleus of a cell that contain the genes.

Clonal life cycle: Growth and development of a plant when propagated vegetatively from a particular propagule of an individual plant.

Clonal propagation: A group of plants originating from a single source plant by vegetative propagation.

Cold frames: Propagation structures covered with poly, lath, or other covering material and which are not heated in the winter.

Conditional dormancy: A continuum seen in many seeds in nature as they cycle through periods of dormancy and nondormancy; it is detected as the seed's ability to germinate over a range of temperature.

Crossing over: The interchange of genetic information during the early stages of meiosis I.

Cultivar: A group of plants that have originated in cultivation are unique and similar in appearance, and whose essential characteristics are maintained during propagation.

Cutting propagation: The clonal multiplication of plants with propagules of stems, leaves, or roots.

***De novo* adventitious roots:** Roots that are formed "anew" (from scratch) from stem or leaf cells that experience a stimulus, such as wounding, to dedifferentiate into roots.

Dedifferentiation: The capacity of mature cells to return to a meristematic condition and develop into a new growing point.

Detached scion graftage: A type of graft used when a section of the shoot of the scion is removed and grafted to the apex or side of the rootstock. It is also used in grafting roots (root graftage).

Domestication: The process of selecting specific kinds of wild plants and adapting them to human use.

Double dormancy: The seeds have dormancy due to hard seed coats and dormant embryos.

Double working: The grafting or budding of an interstock (interstem) between the rootstock and scion.

Edaphic factors: Any factors influenced by the soil or propagation medium (substrate).

Embryo: It represents the new plant generation and develops after the sexual union of the male and female gametes during fertilization.

Endosperm: The major storage tissue in seeds. It is derived from the haploid female gametophyte in gymnosperms, while in angiosperms it is the result of gamete fusion that forms a triploid (3n) storage tissue.

Facultative apomicts: Apomictic species produce both apomictic and sexual embryos on the same plant.

Field nurseries: Nurseries that contain seeds sown at high density in the field for future transplanting to a wider spacing.

Frame working: A form of top budding (topworking) where a few scaffold branches are retained on an established rootstock for multiple budding of a new scion.

Gable roof constructed greenhouse: A unit that has more expensive, reinforced upper support for hanging mist systems, supplementary lights, or additional tiers of potted plants.

Gene hereditary: Unit of inheritance now known to be composed of specific arrangements of nucleotides to make up a genetic code.

Graft chimera: It is arise in grafting at the point of contact between rootstock and scion and will have properties intermediate between those of its "parents". *E.g.*, Bizzaria orange

Graft compatibility: The ability of two different plants, grafted together, to produce a successful union and to develop satisfactorily into one composite or compound plant.

Graft failure: An unsuccessful graft caused by anatomical mismatching, poor craftsmanship, adverse environmental conditions, disease, or graft incompatibility.

Graft incompatibility: An interruption in cambial and vascular continuity leading to a smooth break at the point of the graft union, causing graft failure. It is caused by adverse physiological responses between the grafting partners, disease, or anatomical abnormalities.

Grafting: The union of a root system (understock) with a shoot system (scion) in such a manner that they subsequently grow and develop as one composite (compound) plant.

Hardening off: Hardening off is a term used to refer to the process necessary for a plant to become acclimated to its environment. **Or** it is the process of allowing a plant to transition from a protected indoor or greenhouse environment to the harsh outdoor conditions of fluctuating spring temperatures, wind, and full sun exposure.

Hot frames (hotbeds): Propagation structures that are covered with poly and heated in the winter.

Hybrid: Hybrid plants are the result of cross-pollinating two plant varieties to create a new plant with desired traits.

Intermittent mist: A thin film of water produced through a pressurized irrigation system that cools the atmosphere and leaf surface of cuttings.

Interstock: It is a piece of stem inserted by means of two graft unions between the scion and the rootstock.

Lath houses or shade houses: It provides outdoor shade and protects container grown plants from high summer temperatures and high light irradiance.

Meiosis: The special kind of cell division that results in sex cells, which are utilized in sexual reproduction.

Mericlinal: Occupies only a portion of the outer layer of cells.

Meristem tissue: Tissue composed of undifferentiated cells that can continue to synthesize protoplasm and produce new cells by division.

Meristematic: Cells capable to dividing new cells.

Metaphase: Spindle fibers form and the chromosomes migrate to the center of the cell.

Microclimatic conditions: Any environmental factors (relative humidity, temperature, light, gases, *etc.*) in the immediate vicinity of the propagule during propagation.

Mitosis: The special kind of cell division that results in vegetative propagation.

Monoembryony: It is the phenomenon in which only one embryo produced in a seed.

Morphological dormancy: Such dormancy occurs in some seeds in which the embryo is not fully developed at the time of seed dissemination. Such seeds do not germinate, if planted immediately after harvesting.

Non-recurrent apomixis: The embryo develops directly from the haploid egg cell or from some other haploid cells of embryo sac without fertilization.

Normal seedlings: Seedlings described for the major crops (often in pictures) in the rules for seed testing. In general, normal seedlings have elongated radicle and hypocotyl and at least one enlarged cotyledon.

Obligate apomicts: Some species or individuals produce only apomictic embryos.

Orthodox seeds: Seeds that tolerate maturation drying and survive at less than 10 percent moisture.

Orthotropic: A desirable, upright growth allowing production of symmetrical plant.

Pad and fan system: A system commonly used in greenhouse cooling to reduce the air temperature by raising the relative humidity and circulating air.

Parenchyma cells: It represents the basic living cell type. It is a living cell with a primary cell wall that is metabolically active and capable of differentiating into specific cell types.

Perennials: These are plants that live for more than two years and repeat the vegetative-reproductive cycle annually.

Periclinal: Occupies the outer layer of cells completely.

Physiological dormancy: A condition mainly controlled by factors within the embryo that must change before the seed can germinate.

Plagiotropic: A horizontal branchlike growth habit that is generally not horticulturally desirable.

Plant exchange: The movement of plants from their place of origin to their place of use.

Plugs: Small seedling plants.

Polyembryony: The development of multiple embryos within the same seed.

Polyethylene (poly): A plastic covering used to cover propagation greenhouses.

Propagule: A plant structure used for regenerating plants, which can include cuttings, seeds, grafts, layers, tissue culture explants, and single cells.

Prophase: Chromosomes condense and appear as short, thickened structures with distinctive morphology, size, and number.

Quiescent (quiescence): Buds that are inhibited from growing and elongating via apical dominance of more distal buds produced during the current season on the same shoot.

Quonset type greenhouse: An inexpensive propagation house made of bent tubing or PVC frame that is covered with polyethylene plastic.

Recalcitrant seed: Seed that are unable to withstand maturation drying.

Recurrent apomixis: The embryo sac develops from the diploid egg mother cell or from some other diploid cells of the embryo sac without fertilization.

Repair graftage: Graft used in repairing or reinforcing injured or weak trees.

Rest period: A physiological condition of the buds of many woody perennial species beginning shortly after the buds are formed. While in this condition, they will not expand into flowers or leafy shoots even under suitable growing conditions. After exposure to sufficient chilling hours (1 to 6°C (33 to 43°F), however, the "rest" influence is broken, and the buds will develop normally with the advent of favourable growing temperatures.

Rhizocaline: A hypothetical chemical complex, that was considered important in the biochemical events leading to root initiation.

Rootstock: It is the lower portion of the graft, which develops into the root system of the grafted plant.

Retractable roof greenhouse: It is a modular screen system for greenhouses that helps in saving costs and time along with providing stability, flexibility & durability for the greenhouse structure.

Sectorial: Occupies only a section of the stem extending through all cell layers.

Seed germination: It is process of reactivation of the metabolic activity of the seed embryo, resulting in the emergence of radicle (root) and plumule (shoot).

Seedling life cycle: Growth and development of a plant when propagated from a seed.

Self-incompatibility: It is a form of sexual incompatibility that has evolved to prevent self-pollination within closely related species.

Sexual incompatibility: Genetic trait in which the pollen either fails to grow down the style or does not germinate on the stigma of a plant with the same incompatibility alleles.

Shading: Partial reduction of light to 100 percent light exclusion that can occur during stock plant manipulation and/or propagation.

Species: The natural grouping of plants that have common characteristics in appearance, adaptation, and breeding behavior (*i.e.*, can freely interbreed with each other).

Standard germination: The most common test for seed quality. It is performed according to standards set by seed testing associations, often by certified seed analysts. It represents the percentage of seedlings in a seed lot that germinate normally. In a standard germination test, only seeds that are normal are counted as germinated.

Stratification: A period of moist warm or moistchilling conditions that satisfies dormancy in seeds with endogenous, physiological dormancy.

Telophase: Nuclear envelopes reform around the separated daughter chromosomes.

Thermo dormancy: A type of secondary dormancy that prevents seeds from germinating at high temperature.

Top working: The grafting of a new cultivar onto established trees in the orchard.

Top grafting: A form of grafting in which the shoot of the rootstock is completely removed at the time the graft is made (*e.g.*, in-lay bark graft of pecan).

Totipotency: The capacity of an individual living cell to reconstitute all plant functions.

Vernalization: A period of cold temperature required by plants to induce flowering. In natural systems, these crops grow in late summer, are chilled over winter, and then flower in early spring.

Viability: A measure of whether the seed is alive or can germinate.

Vivipary germination: Seeds germinate prematurely on the plant without desiccation drying or germination of a seed while it is still attached to the mother plant.

Zygote: The life cycle of plants begins with a single cell.

1

Plant Propagation Importance and Scope

Ravi Kumar, Poornita Raturi and M.L. Jat

History of plant propagation

The propagation of plants is one of the absolutely necessary occupations of human beings since time immemorial. Civilization might have started when ancient man learned to plant and grow different kinds of plants, which fulfilled the nutritional and other allied needs of human beings and their animals. Cultivated plants commenced mainly by direct selection from wild species (*e,g.*, lima bean, barley, tomato, and rice), hybridization between species, accompanied by changes in chromosome number (*e,g.*, strawberry, pear, and prunes), and a group of plants occurs naturally as rare monstrosities (*e,g.*, cabbage, broccoli, and brussel's sprouts).

It would have been useless to make advances in plant improvement without methods by which cultivating improved species could be sustained. Unless they are propagated under controlled conditions that recognize their unique characteristics that make them useful, most cultivated plants would either have lost or reverted to less desirable forms.

Theophrastus, a Greek philosopher (circa 300 BC) and disciple of Aristotle, made some of the earliest references to plant propagation. He accounted for many aspects of plant propagation, such as seeds, cuttings, layering, and grafting, in his two books Historia de Plantis and De Causis Plantarum. An example from the translation of De Causis Plantarum illustrates his understanding of propagation: "while all the trees which are propagated by some kind of slip seem to be alike in their fruits to the original tree, those raised from the fruit are nearly all inferior, while some quite lose the character of their kin, as a vine, apple, fig, pomegranate and pear". For example, Pliny recommends soaking cabbage seeds in the juice of houseleek before they are sown to prevent them from being attacked by insects and Columella recommended taking leafless, mallet stem cuttings of grape.

The ever-increasing demand for fruit plants necessitates the need for improved propagation techniques. Tree breeding and propagation techniques serve as an important tool for fruit crop improvement. Fruit plants do not come true to type when raised from seeds, and therefore, these plants are generally raised through vegetative methods of propagation, either by cutting, or layering techniques, or on the roots of other plants (rootstocks) by budding, grafting, or by more progressive methods of propagation *i.e.*, micropropagation techniques. Growing fruit trees on rootstocks instead of their roots has many advantages, which is why most fruit plants are grown on rootstocks. For instance, the invention of glasshouses, mist propagation, hybrid seed production technology, micro-grafting, rooting co-factors in the rooting of cuttings, micropropagation, epicotyl grafting in mango, side veneer grafting in mango, stooling in apple, mango, guava and litchi as well as bottom heat technique in rooting of cuttings are new techniques in the field of propagation of horticultural crops in the last century.

Origin of vegetative propagation

Propagation from cutting began when rooted shoots or suckers were detached and replanted. This led to propagation from un-rooted cuttings. Romans dipped the bases of cutting in ox manure to stimulate rooting. In the Middle East, settlers discovered how to propagate superior forms of grapes, olive, and figs to preserve their desirable characteristics by thrusting woody stems into the soil. By 2000 BC, grafting was fairly common in Greece, the Middle East, Egypt, and China. The original form of grafting was most likely approach grafting because it has a high success rate. The branch of one tree, while still attached to the parent tree, was securely attached to the branch of another tree after the bark of each branch had been wounded. This mimics natural grafting and illustrates how closely people were observing nature. Grafting was used to propagate plants that were hard to root from cuttings and to encourage early fruiting. The roman was among the first to practice detached scion grafting, where a piece of the chosen plant is removed and inserted into a cut in a rootstock with several different fruit cultivars, such as apples, to produce what is now known as multiple trees. The roman and ancient Chinese also employed the technique of budding.

Other natural vegetative reproduction methods were exploited by propagating from food storage organs such as bulbs, tubers, and rhizomes. Plants increased in this way included onion and garlic (Mediterranean), banana (India and Indonesia), pineapple and potato (South America), sugar cane (tropical Africa), and bamboo (Asia).

Simple layering was adapted from the natural layering of wild plants. Records show that the roman was layering grapes in the 1st century BC. Air layering probably began to be used 4000 years ago in China; it is often still referred to as Chinese layering.

Toward the dawn of the first century AD, plant propagation practices were already well-established. Through the centuries that followed, these early propagation techniques were continually developed and improved.

Victorian influences

A concussion of plant-hunting took place in the western world in the 18th and 19th centuries. A wealth of new and existing plants was discovered and traded between Europe and Japan, China, the East Indies, Australia, Asia, Africa, North America, Mexico, and South America. New introductions arrived as seeds, bulbs, or even plants. Eagerness for these new plants and the appeal to grow and propagate them, coupled with the financial wealth of the plant collectors, was the inspiration for the golden age of the greenhouse. Victorians were very inventive in both construction and design. Their methods of controlling temperature and levels of light and humidity in the growing environment of the glasshouse were impressively complex. The greenhouse enabled the creative use of propagation methods and the refinement of techniques. The role of "propagator" becomes important for any garden note. The propagation equipment that was available to Victorian gardeners was fairly primitive compared to modern advances, yet their ideas still form the basis of what is done today. They used cold frames and hotbeds to control temperature and humidity.

Cold frames, sited to capture as much warmth as possible from the sun, especially in winter, were used for seeds, root cuttings, and easy stem cuttings. Bell jars were used in great numbers. The bell-shaped glass jars, about 18 inch (45 cm) tall, were placed over cutting in prepared soil or pots. Although difficult to control precisely, it was possible to maintain high humidity inside the bell's jars. The warmth was provided by solar radiation. Bell jars were effective for raising small quantities of plants from seeds, stem or root cuttings, and even grafted plants. Today, bell jars have largely been replaced by more Versatile & Sculptural Domed Glass. Toward the end of the 19th century, gardeners split the base of a cutting and placed a wheat seed inside the cut stem before inserting the cutting in the soil mix. As the wheat seed absorbed water and began to germinate, it released growth-promoting substances. These helped the cutting root more easily and with more vigour. The practice becomes disappear after 1940 following the introduction of synthetic rooting hormones or auxins. Gardeners also understood the need for seed treatment such as scarification.

Glorious greenhouses

With the advent of the heated greenhouse in Europe during the 18th century, temperature, light, and humidity could be controlled. This extended the range of plants that could be propagated as in this tropical greenhouse (1870).

Modern propagation

Since the 1950s, modern technology and an increase in the shuffle of information among professionals have led to the development of new propagation techniques for the first time in centuries. These new methods, together with modern equipment, make propagation much easier today. Continuing research regularly opens up more possibilities in propagation; these are first tested by professionals and if they prove worthwhile, eventually benefit the gardener.

Mist propagation

In the 1950s, the intermittent mist propagation method was used to root stem cuttings, especially softwood and semi-ripe material. The unit serves bottom heat to stimulate rooting and constant, regulated humidity to keep the cuttings moist and cool. An automatic mist control system is activated when a fall in moisture film levels are detected on the cuttings due to the mist control sensor. Mist propagation is widely used in commercial propagation and is useful for gardeners, if you cannot afford a dedicated unit; create your version with soil warming cables and a misting system in a closed case.

Plastic film

Another development of the 1950s was plastic film. During the cutting process, bottom heat is provided to the cuttings and the plastic film (a clear plastic sheet) is wrapped around them to create a sealed environment. This maintains high humidity around the cuttings' tops. Despite rotting being a problem in cool temperatures; gardeners can easily adopt this system. Plastic film can also be used among cold frames to warm soil before seeds or cuttings are inserted and then to cover new plants in the frame.

Fog propagation

The key development in the mid-1980s was fog propagation, which provides a much smaller water droplet than mist propagation so that the air remains moist for a very much longer period. It also evades wetting the foliage, as in mist propagation, so is perfect for cuttings or seedlings that are prone to rot. In recent years, fog systems have been simplified and made more reliable.

Seed treatments

Seed priming accomplishes the natural ability of some seeds to halt development if soil conditions are unfavourable. It improves the speed and uniformity of germination with a controlled amount of water and then redried just before the radicle (embryonic root) emerges. Timing of the treatment is critical. True germination does not occur until the seeds are germinated or chitted until the seed is sown. In trade, seeds are germinated or chitted, until the radicle emerges, then packed, sometimes in gel, and sent out for immediate sowing. Gardeners can also chit seeds; it is very useful for hard-coated seeds especially vegetables *e,g.*, potato

Pelleted seeds are layered with an inert material, such as a polymer, that splits or softens on contact with water. The coating may contain fungicides nutrients and a fluorescent dye. The pellet makes sowing easier, particularly with small seeds, thus reducing losses.

Micropropagation

This technique is used to propagate a huge number of plants from a small amount of material. By this means, hard-to-produce plants, new cultivars, and virus free crops such as raspberries can be offered to gardeners. To conserve plants in the wild, old and rare plants can be increased from existing stocks. Micropropagation generally involves growing pieces of plant tissue *in-vitro* (in glass) in sterile laboratory conditions. This is possible because of the ability of the most plant to regenerate from a single cell (This topic has also been described in chapter 7).

Tissue culture

Plant tissue culture is rooted in the discovery of cells and the formulation of cell theory. Schleiden and Schwann proposed in 1838 that cells are the basic structural unit of living organisms. As the cell is capable of autonomy, if given the right environment, it should be able to regenerate into a whole plant. This premise led a German physiologist, Gottlieb Haberlandt to culture for the first time single palisade cells from leaves in Knop's salt solution enriched with sucrose in 1902. The cells continued to grow and accumulate starch for up to one month, but were unable to divide. Although he failed, he contributed to the foundation of plant tissue culture technology. After those discoveries of *in-vitro* cell culture, root and stem tips culture, plant growth hormone *e.g.*, Indole acetic acid, vitamin B as a growth supplement in tissue culture media, hormonal control (auxin: cytokinin) of organ formation, *etc.* took place in tissue culture.

Scientific and horticultural literature

Between 300 BC and AD 2, early Greek, Roman, and Arab writers wrote the first significant written works on agriculture, plant medicine, and propagation that shaped western culture. There were undoubtedly many works lost, but many have survived because they were preserved in Arab libraries and passed on through medieval monasteries. As a result of the invention of the printing press in 1436, a resurgence occurred in the production of herbals, which described and illustrated plants with medicinal properties. Most of the information was drawn from older Greek literature from the first century, especially Dioscorides.

Evolutionary genetics and natural selection had a significant impact on science in the late 1800s. Charles Darwin and his origin of species as well as its important contemporary the variation of animals and plants under domestication introduced the concept of evolution and set the stage for the genetic discoveries following the rediscovery of Mendel's papers in 1900. A subsequent explosion of knowledge and application laid the foundation for present-day plant propagation, as well as the increasing understanding of plant growth, physiology, anatomy, and other biological principles. The proliferation of gardening journals and books began. The first book on nurseries, Seminarium was written by Charles Estienne in 1530.

Charles Baltet, a practical nurseryman, published a book in 1821, The Art of Grafting and Budding, describing 180 methods of grafting. A book by Andrew J. Fuller on Propagation of Plants was published in 1885.

Liberty Hyde Bailey published his first edition of The Nursery Book, which was later revised as the Nursery Manual in 1920. This book cataloged what was known about plant propagation and plant production in nurseries at the time. His Cyclopedia of American Horticulture, published in 1900 to 1902, Standard Cyclopedia of Horticulture, published in 1914–1917, Hortus, published in 1930, Hortus Second, published in 1941, and Manual of Cultivated Plants, published in 1940 and 1949, describe the plants in cultivation. An update, Hortus Third, is a classic in the field. M. G. Kains of Pennsylvania State College and, later, Columbia University in New York, published Plant Propagation, Later revised by Kains and McQuesten, which remained a standard text for many years. Several other books were written during this period including titles by Adriance and Brison, Duruz, Hottes, and Mahlstede and Haber. In 1959, Plant Propagation: Principles and Practices was published for the first time. Eight editions have been published since then.

Scope and importance of plant propagation

Horticultural plants, especially fruit trees, are perennials. They can live for about 100 years and produce fruit. Horticulture plays an important role in human nutrition. The crop plays a crucial role in the wealth generation and social economic development of farmers. Nursery units are critical for propagating most horticultural crops vegetatively. There is a wide scope for fruit orchards, ornamental, vegetable, and landscape gardens at public gardens, highways, and cooperative housing societies.

1. When mother plants are obtained, they can easily be propagated by cuttings or root division, reducing propagation costs.
2. In a nursery, seedlings and grafts are produced, as well as fruit trees and ornamental gardens that require the least care, cost, and maintenance.
3. A large number of different species can be multiplied.
4. Plant species that are endangered can be protected.
5. Plants can easily be improved in terms of characteristics and quality.
6. During the planting season, nursery planting materials are available. By raising seedlings in this way, farmers save time, money, and effort.
7. For healthy and vigorous plant growth, genetically pure planting material is essential. Produce quality and healthy plants on a commercial basis.
8. Stock and scion of genetically pure trees are readily available in quantity and quality for further multiplication.
9. The export of nursery stock to other countries has been enhanced as a result of globalization. Special techniques and care are required when exporting nursery stock. Likewise, great care must be taken when importing nursery stock from outside.
10. Grafting, budding, potting, repotting, and other nursery operations require highly skilled professionals. In the nursery, technical, skilled, semi-skilled, and unskilled workers are employed.

Cellular basis for propagation

Genetical concepts about plant propagation

To understand the science of plant propagation properly, one should have three basic features in mind. First, the art of propagation, the knowledge about the plant growth and structure of plant (root, stem, leaves, flower, seed *etc.*) along with knowledge of different types of plants and various possible methods by which different plants can be propagated. For basic knowledge on genetic processes and structure, one should have idea about the inheritance pattern and genetic manipulations that maintains the characters of a plant from generation to generation.

In fact, cell is the basic unit of life and propagation also has cellular base. As indicated earlier, plant propagation involves the control of two different types of life cycle, the sexual (seed) or asexual (vegetative) but both should have the basic capacity to reproduce offspring similar to the mother plant.

Sexual reproduction involves the union of male and female sex cells, the formation of seeds and thereby creation of a population of seeding individuals with new and different genotypes. Meiosis cell division, which also known as reduction division, takes place, which basically involves sex cells and reduction in number of chromosomes to half each in female and male gamete. They unite together, giving rise to new offspring cells. The offspring so produced may resemble either or none of the parents. The inheritance of characters from the parents takes place from generation to generation through the action of gene present on the chromosomes of the cell. Some traits are controlled by single and some by multiple genes. In such cases, two terms, *viz.*, homozygous and heterozygous are used for describing the genotype of a seed or plant produced by it. As stated earlier, a population is usually homozygous if produced by self-pollinated plant and heterozygous if produced by a cross-pollinated plant.

Asexual propagation is possible because each cell of a plant contains all the genes necessary for the growth and development (called as totipotency of a cell) and during cell division (mitosis) these are replicated in daughter cells. In mitosis cell division, the chromosomes divide into two daughter cells. As the chromosomes produced are the same as in mother cells, the daughter cells have all the characteristics similar to the mother cells and ultimately the new plants have the same genotype and phenotype similar to mother plant. Mitosis occurs basically in specific growing plant like shoot apex, root apex. cambium, intercalary zone and callus mass, which results in the growth and development of vegetative parts of a plant. Basically, mitosis is the cell division responsible for wound healing and normal vegetative growth, regeneration, which makes possibility of raising many horticultural plants by cuttings, layering, separation or division. These methods of vegetative propagation are of commercial importance they permit large-scale multiplication of desirable plants. But, what to important is that the plants produced are identical to the mother plants except in some cases, where sudden heritable changes (mutations) or chimeras may alter the genotype and phenotype of the new individuals as well.

Advantages of plant propagation

There are several advantages to sexual propagation, including:

1. Seeds are a simple and easy way to propagate plants.
2. Especially in the selection of chance seedlings, seed propagation is the only means of diversity.
3. These plants are long-lived, productive, and tolerant of adverse soil and climatic conditions as well as disease infestations.
4. Plants like papaya and coconut, for which asexual means of propagation are uncommon, can be propagated from seeds.
5. Hybrids can only be created through sexual reproduction.
6. Asexual propagation could lead to polyembryony (citrus, mango, or jamun) as well as apomixis (*Malus sikkimensis*, *Malus hupehensis*, *Malus sargentii*), which results in true to type plants.
7. For asexual propagation, seed is the source of rootstocks.
8. When stored properly, seeds can be kept for a longer period of time.

Asexual method of propagation has several advantages, like:

1. Asexually propagated plants are similar to their mother plants.
2. An asexually propagated plant has a short growth phase and bears flowers and fruit much earlier than a sexually propagated plant.
3. Vegetatively propagated plants are smaller in stature, so management operations such as spraying, pruning, and harvesting are easier.
4. When seed setting is not possible (*e.g.* pineapple and banana), asexual propagation can be used as a substitute for sexual propagation.
5. Asexual methods make it easy to perpetuate/multiply desirable characteristics from a mother plant.
6. Asexual propagation is usually used to exploit the benefits of rootstocks and scions.
7. Using asexual propagation, such as bridge grafting, it is possible to repair damaged parts of a plant.
8. By using asexual methods, it is possible to change a nonproductive local variety into a productive improved variety.

9. Self incompatibility exists in a number of fruit species like mango, aonla, apple, almond and cherry *etc.*, provision for adequate pollinizer in an established fruit orchard can be obtained by top working with the desirable pollinizer/ cultivar.
10. Viral indexing is a useful technique for studying viruses in citrus. It can be done by grafting or budding the desirable scion on susceptible rootstock (indicator plant) for detecting the presence of tristeza virus in citrus.

Constraint of plant propagation

The sexual method of propagation has some limitations, such as:

1. Fruit plants are heterozygous, which means they are not true to type.
2. Plant seedlings go through a long juvenile phase (6-10 years) and therefore they begin flowering and bearing fruit very late.
3. Generally, sexually raised plants are tall and spreading, making it difficult for them to be pruned, sprayed, harvested, *etc.*
4. Litchi, jamun, jackfruit, citrus, mango, and avocado seeds need to be planted immediately after they are extracted from their fruits because they lose their viability very quickly.
5. Sexual propagation does not exploit the beneficial influences of rootstocks on scion varieties.
6. Fruits from plants are usually inferior in quality.

Asexual method of propagation has some limitations, like:

1 An asexually propagated plant has a shorter lifespan.

2 Diversity is restricted by asexual reproduction.

3 Sometimes, asexual propagation can disseminate diseases, such as tristeza virus in citrus.

4 There is a need for technical expertise/skill.

Opportunities in plant propagation

We know the fact that the nursery industry has problems, but having problems is normal. It changes everything you do when it comes to what you propagate, the way you propagate, how you develop your business, the way you find your way into your industry, *etc.* A key difference today is that change happens much faster, and this is unsettling for many people. Technology, along with better access to information, is not only changing how we work, but it is also continuing to change how we work. For the future, we have only two choices: either embrace change, adapt, and use it to our advantage, or reject change, be overwhelmed, and fail. There will be continued growth and significance of the plant propagation, nursery, and other horticulture industries. We will certainly continue to need to propagate and grow plants. Many of our daily needs are provided by plants, including food, medicine, shelter, and clothing. They are one of our most important tools for controlling environmental issues such as land degradation and climate change. Whenever there is a demand, supply will rise in order to meet it. This is the law of economics. It is uncertain how many existing players in the industry (individuals and organizations) will adapt to future needs; and how many will be replaced by new ones. Technologies such as social media, the internet, mechanization, and scientific advances are offering individuals opportunities they never had before. Career success and failure are largely determined by how one responds to those opportunities.

New scientific discoveries continue to affect plant propagation. The benefits of these techniques are not available to gardeners yet but may be in the future. Recent innovation includes genetic engineering; foreign genes with known, desirable characteristics are transferred into another plant cell. It is possible to introduce a gene that is unrelated to the recipient plant unlike natural hybridizing and traditional selective breeding, both of which also result in offspring that are genetically different from the parent plants. The technology, involving molecular biology, is very complicated and not without problems. An average plant has 20000 different genes, of which there may be five million copies in a single cell, so determining which gene is responsible for which characters can be difficult. The minute scale of the gene transfer operation demands special techniques. The finished cell is micro-propagated to produce a stock plant for propagation.

Genetic engineering has enormous potential to augment the usefulness of existing plants and to create new ones. Current time work is aimed at improving the resistance of crops to disease, cold, and pests. There are concerns, however, about the ramification of introducing plants that could never occur in nature into the environment. Naturally fertilized seeds contain genes from two parents; no two seeds are identical. It is now possible to create artificial seeds (somatic embryos) from vegetative tissue. This comprises isolating embryos grown in solution from single cells and giving them a synthetic coating. A vast number of genetically uniform "seeds" can be produced, which give rise to genetically identical plants.

In micro-grafting, minute pieces of plant tissue are used to produce disease and virus free plants, especially fruit trees. First, the seedling rootstock is raised in sterile conditions. When a seedling reaches the first true leaf stage, it is micro-grafted with the tiny, virus free tip (meristem) of the desired plant. After about six months, micro-grafting is ready for normal planting. Virus free, micro-propagated (clonal) rootstocks may also be used to avoid the variability that can occur with seedling rootstocks.

A glance at the future developments in plant propagation

1. **Better control in climate chambers:** The nursery is acquiring more control over the climate parameters temperature, relative humidity, light intensity, and the photoperiod, and that its transition from mobile tunnels to healing chambers built from insulating panels has meant a reduction in labour.

2. **Better healing and root development with LEDs:** Heat was emitted from fluorescent lamps, and the light spectrum was not optimal for healing. With the switch to LEDs, every aspect of lighting improved: optimum spectrum, low energy consumption, and no heat emissions. Furthermore, plants thrive under LED lamps with the right spectrum, which promotes faster healing. It takes six days for tomato seedlings to heal, and seven days for cucumber seedlings. In addition, the plants have more active roots that develop faster, and when exposed to blue light, the plants remain more compact.

3. **Good microflora:** By encouraging the development of 'good' microorganisms, we can shift the natural balance and reduce *Fusarium's* opportunity to grow. As applied to horticulture, this means that we can stimulate the growth of beneficial microorganisms in the root zone by feeding them purified and disinfected root exudates. A diverse microflora aids plants in taking in nutrients and fending off pathogens from the root environment, resulting in a system that is more resilient.

4. **Latest developments:** Seedlings topped once to the cotyledons and then again to the second leaf, or topped twice to the second leaf, grow into plants with four stems. Growers benefit from these plants due to the fact that they get two to three additional trusses on the third shoot and need fewer plants, and hence fewer seeds per hectare. It is quite a challenge for propagators, since some plant types are not that easy to realize, especially large tomatoes.

5. **Removal of leaves:** Excess leaves and unwanted shoots must be removed to set up a good leaf balance, especially in summer and early autumn crops. Winter crops are delayed by two to three days.

References

Adriance, G.W. and Brison, F.R. 1955. Propagation of Horticultural Plants. New York: McGraw Hill.

Bailey, L.H. 1891. The Nursery Book. Harrisburg, PA: Mount Pleasant Press, J. Horace McFarland.

Bailey, L.H. and Bailey, E.Z. 1941. Hortus Second. New York: Macmillan.

Bailey, L.H. and Bailey, E.Z. 1976. Hortus Third. New York: Macmillan.

Baltet, C. 1910. The Art of Grafting and Budding. 6th ed. London: Crosby Lockwood (quoted by Hottes, 1922).

Collins, J.L. 1960. The Pineapple-History, Cultivation, Utilization. London: Leonard-Hill.

Cunningham, I.S. 1984. Frank N. Meyer: Plant Hunter in Asia. Ames, IA: Iowa State University Press.

Darwin, C. 1859. The Origin of Species by Means of Natural Selection, or the Preservation of Favoured Races in the Struggle for Life. London: J. Murray.

Davidson, H., Mecklenburg, R. and Peterson, C. 2000. Nursery Management. 4th ed. Upper Saddle River, NJ: Prentice Hall.

Duruz, W. P. 1949. The Principles of Nursery Management. 1th ed. New York: A. T. de la Mare Co.

Duruz, W. P. 1953. The Principles of Nursery Management. 2nd ed. New York: A. T. de la Mare Co.

Evenari, M. 1981. The history of germination research and the lesson it contains for today. *Israel J. Bot.*, 29: 4-21.

Gautheret, R.J. 1983. Plant Tissue Culture: A History. *Bot. Mag.*, 96: 393-410.

Goldschmidt, 2009. A History of Grafting. *Hort. Rev.*, 35: 437-87.

Hartmann, H.T. and Kester, D.E. 1959. Plant Propagation: Principles and Practices. Englewood Cliffs, NJ: Prentice-Hall.

Hartmann, H.T., Kester, D.E., Davies, F.T. and Geneve, R.L. 1997. Plant Propagation: Principles and Practices. Prentice Hall of India Private Limited, New Delhi. p.770.

Hartmann, H.T., Kofranek, A.M., Rubatsky, V.E. and Flocker, W.J. 1988. Plant Science: Growth, Development and Utilization of Cultivated Plants. 2nd ed. Englewood Cliffs, NJ: Prentice Hall.

Hendrick, U. P. 1933. A History of Agriculture in the State of New York. New York: Hill and Wang.

http://ecoursesonline.iasri.res.in/mod/page/view.php?id=96779

https://www.acsgarden.com/articles/careers-advice/future-of-propagation.aspx

Mahlstede, J.P. and Haber, E.S. 1957. Plant Propagation. New York: Wiley. 29.

Malpighi, M. 1675. Anatome Plantarum. London.

McDonald, M.B. 1994. The history of seed vigor testing. *J. Seed Tech.* 17:93–101.

Mendel, K. 1994. The history of plant propagation methods during the last 70 years. *Acta Hort.*, 314: 19-26.

Michael J.D.M. 2010. A history of Hawaiian plant propagation. *Sibbaldia,* 8: 31-43.

Mudge, K., Janick, J., Scofield, S. and Goldschmidt, E.E. 2009. A history of grafting. *Hort. Rev.*, 35: 437-493.

Read, P.E. 2007. Micropropagation: Past, present and future. *Acta Hort.*, 748: 17-27.

Reed, H.S. 1942. A Short History of the Plant Sciences. New York: The Ronald Press Co.

Reuther, G. 1990. Current status and future prospects of large scale micropropagation in commercial plant production. *Food Biotechnol.*, 4(1): 445-459

Sauer, C.O. 1969. Agricultural Origins and Dispersal. 2nd ed. Cambridge, MA: Massachusetts Institute of Technology Press.

Sharma, R. R. 2002. Propagation of Horticultural Crops: Principles and Practices. Kalyani Publishers, New Delhi.

Teresa, M., Rodrigues, M. and Gimeno, L. 2015. Constraints propagation techniques in batch plants planning and scheduling. *European Symposium on Computer Aided Process Engineering*, 15: 1225-1231.

Toogood, A. 1999. American Horticulture Society Plant Propagation. D.K. Publishing, INC. 8-46.

Kubota, C., McClure, M. A., Kokalis-Burelle, N., Bausher, M. G. and Rosskopf, E N. 2008. Vegetable grafting: History, use, and current technology status in North America. *Hort. Science*, 43 (6):1664–69.

Wells, J.S. 1981. A history of rhododendron cutting propagation. *Amer. Nurs.* 154:14-5.

Zohary, D. and Spiegel-Roy, P. 1975. Beginnings of fruit growing in the old world. *Science*, 187(4174): 319-27.

2

Apomixis, Polyembryony and Chimeras

Sherya and R.K. Jat

Introduction

Apomixis is a phenomenon in which there is development of seed without fertilization or sexual fusion. Embryo develops without fertilization in apomixis. Hence, apomixis is an asexual means of reproduction therefore the genotype of plant will be same as of mother plant. Apomixis was first time discovered by Leuwenhock in 1719 in citrus whereas the term apomixis was first time used by Winker in 1908.The main benefit of apomixis over sexual reproduction is the selection of individual plant with superior characters and propagation of clonally through seeds. Higher plants commonly exhibit apomixis. There are more than 300 species belonging to 35 families that exhibit apomixis. Most commonly, it occurs in Gramineae, Compositae, Rosaceae, and Rutaceae.

The plant species, in which this phenomenon is found, called as apomictic species. In apomictic species sexual reproduction is either absent or suppressed. Sometimes, sexual reproduction is also reported in plants along with apomixis known as facultative apomixis. But when sexual reproduction is absent is called as obligate apomixis.

Types of apomixis: It is classified into four categories: (1) Parthenogenesis (2) Apogamy (3) Apospory and (4) Adventive embryony.

(1) Parthenogenesis: This is the process of developing embryos from egg cells without fertilization. It is of two types *i.e.,* (a) haploid parthenogenesis (when haploid egg cell give rise to the embryo). Plants which are developed from such embryo are haploid and sterile. (b) diploid parthenogenesis: Sometimes embryo sac develops without reduction division. Such embryo sac and all the cells within it are diploid and give rise to diploid embryo. *E.g.,* Mangosteen (*Garcinia mangostana*). There are four main ways in which parthenogenesis can be induced artificially: (1) by the stimulation of widely related pollen or foreign pollen, (2) by low temperature, (3) by pollinating with X-rays irradiated pollens (4) by treatment with certain chemicals like belviton. They all promote the development of the parthenogenetic process in egg cells.

(2) Apogamy: When synergids or antipodal cells of the embryo sac give rise to the embryo is called as apogamy. There are two types of apogamy (a) haploid apogamy and (b) diploid apogamy. Synergides and antipodal cells can be both haploid and diploid. In haploid apogamy, embryos are formed from haploid cells and in diploid apogamy, embryos are formed from diploid cells.

(3) Apospory: During this process, the first diploid cell of an ovule that lies outside an embryo sac develops into a second embryo sac without reduction. Consequently, the embryo develops directly from the diploid egg cell without fertilization. Two types of apospory exist: (a) generative apospory (b) somatic apospory. generative apospory refers to the development of embryos from cells of archesporium. While as a result of somatic apospory, embryos develop from cells of the nucellus or integument as the embryo sac. This occurs in *Malus*.

(4) Adventive embryony: The formation of embryo directly from the diploid cells of the ovule outside the embryo sac which are either in the nucellus or integument is termed adventive embryony. Sporophytic budding is a common feature in citrus and mango trees. *E.g.*, mango cvs. Olour, Goa, Kurukkan, Bappakai, Vellaikolamban, Nileswar Dwarf, Salem, Bellary, Goakasargod, Mazagaon, Chandrakaran *etc.* and most of the species of citrus except *Citrus medica* (Citron), *Citrus grandis* (Pummelo or Shaddock) and *Citrus latifolia* (Tahiti lime).

Apomixis can also be classified as recurrent (2n) or non-recurrent (n). During recurrent apomixis, the embryo sac develops from diploid cells. The number of chromosomes in the embryo do not decrease and all embryo sac cells formation from diploid cells. Since this process occurs from generation to generation, it is called recurrent apomixis. Recurrent apomixis involves diploid parthenogenesis, diploid apogamy, and diploid apospory. Recurrent apomixis occurs most often in the following species: *Parthenium*, *Rubus*, *Malus*, *Allium*, *Rudbeckia*, *Poa*, *Taraxacum etc.*

In non-recurrent apomixis, the embryo sac consists of usual haploid cells. It produces plants that contain haploid chromosomes and are usually sterile. As this process does not repeat from generation to generation, it is known as non-recurrent apomixis. The process involves haploid parthenogenesis and haploid apogamy. It occurs only in a few species, such as *Solanum nigrum*, *Lilium* spp. *etc.*

Exploitation of apomixis in crop improvement: It is very important to detect or identify the naturally occurring phenomenon of apomixis for its exploitation in sexually reproducing crops. Possibly, this artificial incorporation could occur by means of hybridization between apomicts and amphimicts.

Detection of apomixis: Evidence for apomixis is obtained only after a thorough screening of many different varieties and hybrids. It involves tracing carefully and systematically the stages for the development of embryo sac and embryos, starting with microtomy of the ovule and proceeding right to embryonic development. Because of this, it requires patience and persistence.

It must still be noted that plant breeding benefits only from recurrent apomixis, namely diploid apospory, parthenogenesis, apogamy, adventitious embryony and vegetative propagation. This is because these diploid forms produce viable diploids without fertilization and can continue to reproduce for generations without requiring re-fertilization. Non-recurrent apomixis is of academic importance only.

Maintenance and transfer of apomixis: When an apomict plant is detected, its inheritance pattern may be studied by crossing a few flower samples with pollen obtained from normal plants and observing the segregation patterns of F_2 and subsequent generations. The remaining flowers may thoroughly be checked and seeds collected on maturity. The true apomictic plant will automatically produce mother apomictic progenies, which can be easily maintained.

For many apomictes, there is evidence of hybrid origin in regard to transfer of apomixis. However, there is no proof that hybridization can, by itself, induce apomixis. Situation is further aggravated by the unstable nature of apomicts since there is likelihood of the breaking down of interacting gene complexes conditioning apomixis. Non-apomicts have the least chance of being infected with apomixis, but they aren't totally excluded.

Advantages of apomixis: With the help of apomixis faster reproduction of genetically identical individuals can be done without the problem of segregation, and hybrid vigor or heterosis can be fixed permanently in plants. As well, apomixis may also affect an efficient exploitation of maternal influence, if it occurs, in the resulting progenies, early or delayed, since it perpetuates only specific maternal properties due to the prohibition of fertilization. Horticultural crops, particularly fruit trees and ornamental plants, are most vulnerable to maternal effects.

Disadvantages of apomixis: Apomixis can be a nuisance for breeders when they are trying to produce sexual offspring.

Polyembrony

More than one embryo exists within a single seed that is phenomenon known as polyembryony. That is why it leads to multiple seedlings emerging from one seed. These multiple seedlings are formed from other maternal tissue of ovule along with the zygote. Previously, this phenomenon of polyembryony was considered as unwanted process but now a days it is believed to be very desirable in case of certain plants like mango, citrus, jamun, rose apple *etc.* to obtain true to type seedlings.

Braun studied this phenomenon for the first time in 1859 in plants; Leeuwenhoek reported it for the first time in citrus in 1719. The formation of multiple embryos was first time demonstrated by Strasburger in 1878 in angiosperm. An estimated 250 species of 150 genera from 63 angiosperm families exhibit polyembryony.

There are so many ways through which polyembryonic induction can happen in plants. Most commonly, sporophytic cells of the ovule are activated when an embryo is formed. When embryo arising from maternal sporophytic cells outside the embryo sac are called adventive embryos. In case of most of horticultural crops nucellar embryony is the cause of polyembryony.

The number of embryos per seed varies from three to five in rough lemon. Citrus species are polyembryonic except *Citrus medica* (Citron), *Citrus latifolia* (Tahiti lime) and *Citrus grandis* (Pummelo) which are monoembryonic. In citrus, nucellar embryony is beneficial for producing plants that are vigorous, uniform, and virus free. However, it is a barrier to hybridization. A vigorous growth of nucellar embryos inhibits the zygotic embryo's growth and causes its degeneration prior to seed maturation in polyembryonic cultivars. It is possible to restore such aberrant embryos by tissue culture. The number of embryos varies from 2 to 10 in certain cultivars of polyembryonic mangoes. Polyembryonic seedlings are distinguished from their sexual counterparts by their uniformity and vigorous growth. Polyembryony in mango is determined by a single dominant gene.

Classification

On the basis genetic composition, polyembryony can be classified in two groups *viz.,* Sporophytic: When multiple diploid embryos arise either from zygote or nucellus/intgument (sporphytic cells of ovule) is called as sporophytic polyembryony. In a gametophytic polyembryony, multiple embryos are formed from the gametic cells of the embryo sac (synergid, antipodal) with or without fertilization. In this case haploid embryos are formed.

Horticultural crops jamun, citrus, rose apple, mango, almond, peach *etc.* are polyembryonic in nature. This behavior is very much prominent in citrus group of species. All species of citrus are polyembryonic except *Citrus medica* (Citron), *Citrus latifolia* (Tahiti lime) and *Citrus grandis* (Pummelo). The occurrence of polyembryony in mango is variety dependent. Polyembryonic varieties of mango are Bappakai, Chandrakaran, Kensington, Kitchner, Kurukkan, Muvandan, Mylepelian, Nekkare, Olour, Peach, Prior and Starch. Generally, polyembryonic varieties are used as rootstock due to inferior fruit qualities.

Identification of nucellar seedlings

In the case of facultative polyembryony, there should be certain criteria to be followed to distinguish between polyembryonic seedlings from the zygotic one. Moreover, this identification criteria varies from crop to crop. Sometimes morphological markers are helpful in the identification of it like in case of citrus group by crossing the parent with other having trifoliate leaves. These trifoliate leaves appear in the zygotic seedling and distinguish it from nucellar embryo. In case of *Mangifera indica*, the sexual seedlings are more vigorous and possess giant cotyledon. Even though this visual distinction is not the effective. Sometimes, the nucellar seedlings arise at the lateral position with respect to the zygotic seedling. Generally, zygotic seedlings are weaker and less vigorous as compared to the polyembryonic one. Therefore, biochemical and molecular markers (RAPD, RFLP, SSR *etc.*) are presently used to distinguish sexual seedling from nucellar seedling at an early stage of seedling development.

Importance of polyembryony

- Nucellar seedling provides better rootstock which directly influences the yield of fruit crops.
- Polyembryony has been observed as a mechanism to restore vigor after recurrent asexual reproduction.
- It increases the adaptability of seedlings under adverse climatic condition.
- In case of citrus, it can be used to produce plants that are virus free.

Chimeras

Chimera is a condition in which an individual contains different genotypes. The presence of different genotypes in the same individual is the result of somatic mutation. If the formation of chimera is useful it can be perpetuated asexually.

First time this term chimera was used for the genetic variation arisen in clonal population. It arises due to the distinctive style of apical meristem and fixed location of mutation within the diving cell near the apex of the apical meristem. A chemical may be used to induce somatic mutations at the terminal or axillary buds of germinating seeds, seedlings, or mature plants. Such treatment of buds, seeds, vegetative propagules can create chimeras which is having very significant role in the creation of new horticultural and agricultural crops. The meristematic tissue which is present at the shoot tip contains two functional layers *viz.,* outer and inner layers. The outer layer contains epidermis and part of the mesophyll cells, while the inner layer contains the rest of the tissues, including the reproductive organs.

There are three types of chimeras: (1) Periclinal chimera: When a chimera affects both the outer and inner layers, this is known as a 'periclinal chimera' (inner periclinal or outer periclinal depending upon the layer affected), (2) Sectorial chimera: Only a part of the inner or the outer layer is affected and (3) Mericlinal chimera: In mericlinal chimera, the combination is similar to the periclinal except that the cells carrying the mutant genes occupy only a part of the outer cell layer (As represented in the figure below).

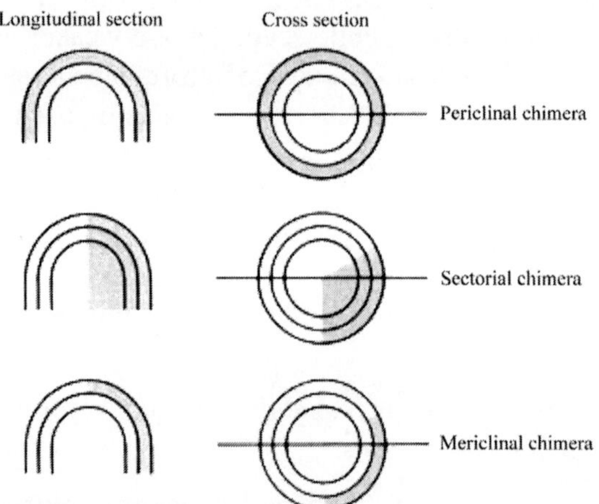

With clonal crops, both outer and inner chimeras can be used, whereas with sexually reproducing species only inner chimeras (periclinal or sectorial) will be transmitted to the next generation. Sectorial chimeras do not survive in clonal crops, so they must be made periclinal through successive clonal propagation and selection.

References

Kumar, N. 2006. Breeding of Horticultural Crops: Principles and Practices. 1st edition, New India Publishing Agency, New Delhi. pp.38-42.

Shukla, A. K., Shukla, A. K. and Vashistha, B. B. 2004. Fruit Breeding: Approaches and Achievements. 1st edition, New India Publishing Agency, New Delhi. pp.4-6.

Singh, P. 2006. Essentials of Plant Breeding. 3rd edition (Revised), Kalyani Publishers, New Delhi. pp.53-55.

3

Seed Dormancy and Germination

Shubham Jagga

"The light of unconditional love awakens the dormant seed potential of the soul helping them ripen, blossom and bear fruit, allowing us to bring the unique gifts that are ours to offer in this life"

–**John Wehwood**

Introduction

In nature and in humans, seeds are vehicles for spreading new life from place to place. In plants, it refers to the unit of dispersal that survives between seed maturation and its establishment as a seedling. A seed, like a fortress, protects and sustains life as they are well-stocked with special food supplies to endure long sieges. There are many things that seeds are, but above all else, they are the means of survival of their species. It is a way of suspending embryonic development and then resuming it years later after the parents have passed away. Significance of good seed is also mentioned in oldest texts like Rigveda as "Subeejam Sushetre Jayate Sampaghte" *i.e.*, good quality seed sown in good field yields in plenty.

Seed refers to the fertilized, mature ovule containing an embryonic plant, stored material and a protective coat. Seeds are extremely variable in shape, size, colour and length of time during which they remain viable. With all this variations, true seeds are known to have three essential parts in common which actually serve as their distinguishing characteristics:

a) **Embryo:** The most vital component of a seed. A living part that grows out of fertilized egg cells after seed maturity and is restricted by seed maturity. Its components are the radicle, the plumule, and one or two cotyledons.

b) **Stored food:** It is another component of seeds which is deposited while they are still attached to the mother plant. The reserve food may be contained in embryo as in peach or may be in endosperm which closes embryo of certain kinds of seeds like onion. In seed the principal kinds of food stored are chiefly carbohydrates (such as starch, hemicellulose and sugars), fats and proteins.

c) **Testa:** The outer covering that forms the seed's protective coat. It is formed normally from two integuments of the ovule, or in some seed from a single integument.

Types of seeds

Based on the extent/longevity of storage of fruit seeds can be classified into three types which are explained as under (Roberts, 1973).

a) **Orthodox seed:** This type of seed is dried down to around 5-10 % moisture content and can survive at a low or sub-freezing temperature. These seeds remain viable in a dry state for a predictable period of time. During *ex-situ* conservation, these seeds survive drying and freezing. Notable fruit crops include guava, sapota, banana, apple, cherry *etc.*

b) **Recalcitrant seeds:** Seeds that cannot be dried below high relative moisture content (30-50 %) and cannot be stored for an extended period of time. Their vitality period is short and they are sensitive to desiccation. Recalcitrant seeds are those that cannot be dried and frozen during *ex-situ* storage. *E.g.*, jamun, litchi, mango, mangosteen, durian, citrus, avocado, rambutan *etc.*

c) **Intermediate seeds:** Intermediate seeds have the same drying tolerance as orthodox seeds; however they are more sensitive to low temperatures than recalcitrant seeds. *E.g.*, papaya, macadamia nut.

Seed germination

The process whereby the embryo resumes growth and the radicle and plumule break through the seed coat is known as seed germination. It is defined as the process of water absorption by a quiescent dry seed that leads to the extension of an embryonic axis (Bewley and Black, 1994). A seed's germination process is triphasic, *i.e.*, it comprises three steps: phase I rapid initial uptake; phase II plateau phase and in phase III further increase of water uptake, however, only when germination occur (Manz *et al.*, 2005).

Types of seed germination

a) **Epigeal germination:** It is the process by which the cotyledons emerge from the soil and turn green. In dicots, they are pushed up by rapid growth of the hypocotyl before the growth of the epicotyl. In this case straightening of hypocotyl raises the cotyledons and shoot apex towards light. *E.g.,* bean, tamarind *etc.*

b) **Hypogeal germination:** In this germination type, the cotyledons remain underground while hypocotyl growth is restricted. Epicotyl can grow by raising the first leaves from the soil. *E.g.*, custard apple, pea, gram, mango *etc.*

c) **Viviparous germination:** This is a special type of germination which usually occurs in mangroves. During this type of germination, the seed is still attached to the parent plant as it germinates. The embryo emerges from the fruit with an immense radicle pointing downwards, and because of its increased weight, the seedling separates from the parent plant and settles in muddy soil.

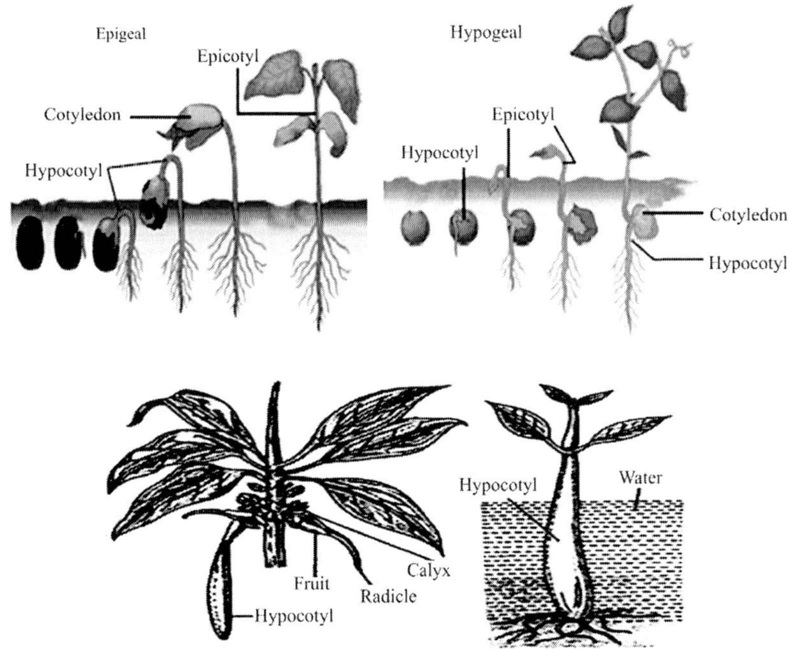

Viviparous germination

Process of seed germination

For the initiation of germination, it is essential that seed is subjected to favourable supply of moisture. During germination the embryo, and the endosperm if present, swell and push off the seed coat which is necessary for processes of gaseous interchange, particularly the exchange of oxygen and carbon dioxide. The seeds contain a relatively large quantity of stored food in form of insoluble forms as starch, fat and protein which must be transformed into soluble compounds in order to be used by seed. Certain specific enzymes, produced within cells carry this process and render the stored foods available. Food then moves to growing parts of radicle and plumule. At the tips of radicle and plumule, new cells and tissues are being formed and elongation of others in process. Following this next stage consists of cell division. Here fresh and dry weight of seedling increases but wet weight of storage tissue decreases.

Factors influencing seed germination of fruit crops

1. **Moisture:** Seeds require a certain amount of moisture to fully saturate and soften. It is required for seed germination. The absorbed water hydrates the cell walls and starch grains, fills the embryo and all empty spaces in the seed, and saturates all the living cells. The seed volume increases with this uptake of water.

2. **Temperature:** Temperature has an effect on germination by its influence on the rate of water intake and the speed of metabolic processes within seed. Maximum and minimum temperatures are the highest and lowest respectively, at which germination will take place and these provides the extreme between which germination will occur for given plants. The temperature within the range at which germination will take place within shortest time is known as optimum. A higher percentage of germination is likely to be obtained at the optimum temperature, and plants produced are usually strong and sturdy.

   Example. Peach and plum germinate within temperature range of 36-90^0 F, for apples, pear and grapes; a temperature of 70^0 F is optimum. Citrus and avocado seeds germinate satisfactorily at about 80^0 F.

3. **Light**: Seeds can be classified into different types upon the light on germination but seeds which are indifferent to light and apparently germinate equally well in light or darkness is true for most fruit crops like pecans, peach, blackberry *etc.*

4. **Oxygen:** The embryo requires oxygen for respiration as well as for initiation of growth. In compact, poorly prepared or excessive wet seed beds germination may be retarded or prevented by lack of oxygen.

5. **Vigour of parent plant:** Seeds from weak plants are deficient in stored foods and thus are known to possess small embryos and will produce less vigorous seedlings than those from normal plants.

6. **Age of seed:** The vitality (vigour or strength possessed by seeds for growth) of seeds of different species is influenced by varying degrees by age. Tropical plant seeds remain viable for short period of time.
7. **Storage of seeds:** Different kinds of seed remain viable longest and ultimately germinate best if stored under conditions especially suited to their specific requirements. Seeds of most fleshy fruits remain viable longest if they are kept moist and cool during storage. Also, if seeds of certain dry fruits likewise lose their viability if allowed to become dry for long period.

Seed dormancy and quiescence

Lang (1987) gave most accepted terminology of dormancy. According to this definition, dormancy is temporary suspension of growth of an organ/tissue having meristematic activity.

A condition in which seeds fail to germinate, even under favourable environmental conditions for germination. These conditions are caused by the fact that seeds require a period of rest before they are able to germinate. The period of rest can range from days to months or even years. However, it differs from quiescence, where the seeds would not germinate or produce seedlings due to unfavourable external conditions, like moisture, temperature, *etc.* It has been observed that some fruit plants (mango, citrus) germinate immediately after being extracted from the fruit under favourable conditions of moisture, temperature, and aeration. However, in other plants (apple, pear, cherry), germination does not take place despite favourable conditions due to dormancy.

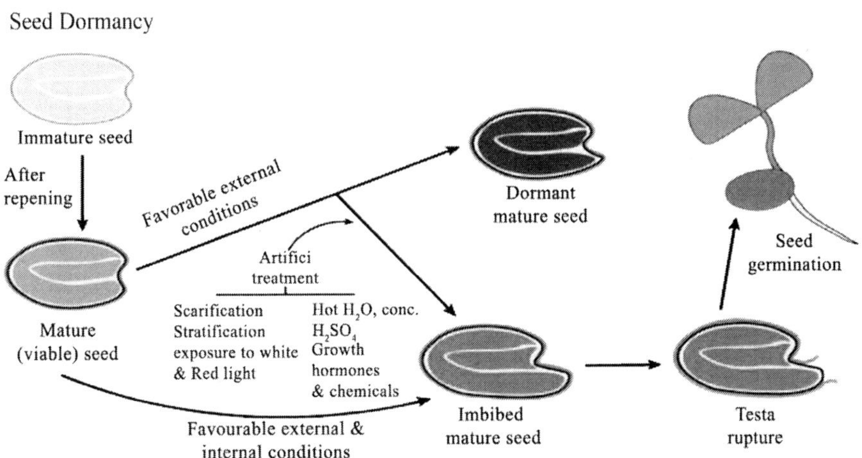

Types of dormancy

1. Exogenous dormancy

This type of dormancy is imposed by external factors outside the embryo. During exogenous dormancy, the tissues enclosing the embryo play an important role in germination by inhibiting water uptake, restricting embryo expansion and radicle emergence, modifying gaseous exchange (limiting oxygen consumption), preventing leaching of inhibitor from the embryo and supplying inhibitor to the embryo. It is further of three types:

a) **Physical dormancy (seed coat dormancy):** In the case of physical dormancy the seed coat or seed covering may become hard, fibrous or mucilaginous (adhesives gum) during dehydration and ripening as a result they become impermeable to water and gases, which prevents the physiological processes initiating germination. Drupes fruits exhibit this type of dormancy commonly, *i.e.*, olive, peach, plum, apricot, cherry *etc.* (hardened endocarp), walnut and pecan nut (surrounding shell).

b) **Mechanical dormancy:** Some fruits have covered seeds that prevent radical growth, resulting in dormancy. During germination, some seed coverings, such as walnut shells, stone fruit pits, and olive stones, are too strong to allow dormant embryos to expand.

c) **Chemical dormancy:** Some seeds of certain fruits are protected from germination by chemicals that accumulate in fruit and seed tissue during development and persist after harvest. It mainly occurs in fleshy fruits or fruits with seeds that remain in their juice, such as citrus, stone fruits, pears, and grapes. There are several phenols, coumarins, and abscisic acids associated with inhibition. Consequently, seeds are inhibited from germinating.

2. Endogenous dormancy

A dormancy of this nature is induced by a rudimentary or undeveloped embryo at the time of maturation or ripening. There are several types of dormancy, including morphological, physiological, double dormancy, and secondary dormancy.

A. Morphological dormancy (rudimentary and linear embryo): Dormancy occurs in some seeds in which the embryo is not fully developed at the time of seed dissemination. Such seeds do not germinate, if planted immediately after harvesting. Plants with rudimentary embryos produce seeds with little more than a pro-embryo embedded in a massive endosperm at the time of fruit maturation. Enlargement of the embryo occurs after the seeds have imbibed water but, before germination begins. Formation of rudimentary embryo is common in various plant families such as Ranunculaceae (Ranunculus), Papavaraceae (poppy).

B. Physiological dormancy

a) **Non-deep physiological dormancy:** In dry storage, seeds must lose dormancy within a set period of time after ripening. The onset of this type of dormancy is often temporary, and it disappears after dry storage. Fruits such as apple, pear, cherry, peach, plum, and apricot have physiological dormancy, which may last from one to six months and disappear with dry storage.

b) **Photo dormancy:** A seed that needs either light or darkness to germinate is called a photodormant seed. Several plants contain phytochrome pigments that are photochemically reactive. The red form of phytochrome (Pfr) is formed as soon as imbibed seeds are exposed to light (660-760 nm), thereby initiating the germination process. In contrast, far-red light (760-800 nm) causes Pfr to change to Pf, which inhibits germination.

c) **Thermo dormancy:** Some seeds require a specific temperature to germinate, otherwise they remain dormant. Such seeds are called thermo dormant seeds. Physiological dormancy can be classified into three groups:

i) **Intermediate physiological dormancy:** Certain species require a minimum of 1-3 months of moist chilling in an aerated and imbibed state, known as moist chilling. For example, the seed dormancy of most temperate fruit seeds can be overcome by moist chilling. Due to this requirement, horticultural stratification has become a world-famous practice. During the process involves placing moist sand between layers of seeds in boxes, which are then exposed to temperatures of 2 to 7°C for up to 3-6 months to overcome dormancy.

ii) **Deep physiological dormancy:** Seeds that usually require a long time (>8 weeks) of moist chilling stratification to remove dormancy, as in peaches.

iii) **Epicotyl dormancy:** It refers to seeds with a separate dormancy condition for the epicotyl and radical hypocotyl. It is known as epicotyl dormancy.

C. Double dormancy: Seeds of some species are dormant because of hard seed coats and dormant embryos. In nature, such seeds take two years to break dormancy. During the first spring, microorganisms make the seeds weak and brittle so that there is an embryo dormancy that is broken by a cold winter next year. The occurrence of more than one type of dormancy is known as double dormancy. It can be morpho-physiological *i.e.*, combination of under developed embryo and physiological dormancy or exo-endodormancy *i.e.*, combination of exogenous and endogenous dormancy conditions *i.e.*, hard seed coat (physical plus intermediate physiological dormancy).

D. Secondary dormancy: This occurs as a result of germination conditions. An imbibed seed is not allowed to germinate if the surrounding conditions are not favourable. These conditions can include high temperatures, low light, and prolonged darkness, as well as stress related to water. It can be divided into three types:

i. **Thermo dormancy:** Dormancy resulting from high temperatures.

ii. **Photo dormancy:** Seeds exposed to excess light for a long period of time.

iii. **Skotodormancy:** Germination requires light when they are soaked in darkness for an extended period of time.

Methods to break dormancy

The dormancy of fruit seeds can be broken using different methods. These are briefly described here under:

a) **Scarification:** This is a method for making seed coverings permeable to gases and water by scratching, breaking, or mechanically altering them. Scarification may be accomplished through mechanical, chemical, hot water, and warm moist methods.

   i) **Mechanical scarification:** It is a simple process that involves chipping the hard seed coat with sandpaper, cutting it with a file, or cracking it with a hammer for relatively small seeds.

   ii) **Acid scarification:** Dry seeds are placed in containers and covered with concentrated sulphuric acid (H_2SO_4) or hydrogen chloride (HCl) in a ratio of one part seed to two parts acid. Depending on the species, the time may vary from 10 minutes to 6 hours. Afterward, the acid is poured off and the seeds are washed. It can either be planted immediately when wet or it can be dried and then planted at a later time.

   iii) **Hot water scarification:** In this process, seeds are immersed in hot water varying from 77° C to 100° C temperature. As soon as the heat source is removed, the seeds are soaked in the slowly boiling water for 12 to 24 hours and immediately planted after treatment.

iv) **Warm moist scarification:** The seeds are placed in moist warm medium for many months to soften the seed coat and other seed coverings through microbial activity. The treatment is especially helpful to double-dormant seeds. In the summer or the fall, stone fruits (cherry, plum, apricot, peach) germinate better if planted early enough to provide one to two months of warm temperatures before frost sets in.

b) **Stratification:** It is the process of rehydrating dormant seed in which a period of chilling is applied to the imbibed seeds to ripen the embryo in alternate layers of sand or soil for a period of time. It is also known as moist chilling. Many tropical and subtropical species (like palms) require the embryo to be subjected to a warm stratification period prior to germination, whereas seeds with hard endocarps, such as *Prunus* spp. (the stone fruit including cherry, plum, apricot and peaches) show increased germination if planted early in the summer or fall to provide one to two months of warm temperature prior to the onset of chilling.

c) **Leaching of inhibitors:** It is well known that phenolic compounds and some inhibitors are present in seed coverings of several species, causing inhibition of the germination process. In this regard, soaking seeds 12-24 hours in water or placing them in water for few hours helps in leaching off inhibitors and phenolic compounds, which aid in easy seed germination.

d) **Pre-chilling:** In some plants, the dormancy of seeds can be overcome through pre-chilling. The soaked or imbibed seeds are kept at a temperature of 5-10° C for 5-7 days before sowing in this treatment. Afterwards, seeds can be immediately sown in the field.

e) **Pre-drying:** This technique also helps overcome seeds dormancy. For this treatment, the dry seeds are heated to 37-40°C for 5-7 days before sowing. After this, seed can be sown in the field.

f) **Seed priming:** Seed priming refers to the procedure of overcoming dormancy in newly harvested fruits. Among the most common seed priming procedures are osmo-conditioning, infusion, and fluid drilling. In osmo-conditioning, the seeds are placed in shallow layer in a container having 20-30 per cent solution of polyglycol (PEG). The seeds are then incubated at 15-20°C for 7-21 days, depending on the seed size and species. During infusion, hormones, fungicides, or insecticides, as well as antidotes, are infused directly into dormant seeds by organic formulations. This process involves soaking seeds for 1-4 hours in acetone or dichloromethane solution containing chemicals. During fluid drilling, the seeds are suspended in a gel before they are sown. Today, there are many types of gels available in the market, but sodium alginate, guar gum, and synthetic clay are the most common materials used in fluid drilling.

g) **Treatment with chemicals:** Some chemical compounds are also used to break dormancy, but their purpose is unclear. For example, thiourea has been reported to stimulate germination in some types of dormant seeds. Soak the seeds in a solution of thiourea containing 0.5 to 3 % for 3 to 5 minutes. Thereafter, seeds are rinsing with water before being sown in the field. Likewise, potassium nitrate and sodium hypochlorite also stimulate seed germination in some plant species.

h) **Hormonal treatment:** GA_3 is one of several hormones that are commercially used to break seeds dormant. In general, a GA_3 concentration of 200-500 ppm is most commonly used, depending on the type of seed. Cytokinins are another group of hormones that break physiological dormancy and stimulate plant germination. Commercial preparations of cytokinin, like Kinetin and BA (6-benzyl aminopurine), are used to break dormancy in seeds. For many species, soaking seeds in 100 ppm of kinetin solution for 3-5 minutes is highly effective for breaking dormancy. In some species, etheral also stimulates germination.

Hormonal control of dormancy

The seed dormancy and germination processes are mediated by endogenous hormones which may be growth promoting or inhibiting. The commonly associated growth promoting hormones are gibberellins, cytokinins and ethylene, whereas abscisic acid acts as inhibitor. Hormonal application has been proven to be most successful aid in germination process of many seeds. The success, however, depends upon species, variety and the stage of dormancy or germination.

a) **Gibberellins:** Among different hormones, gibberellins are directly related with dormancy or germination of seeds. Usually, these compounds are concentrated in developing seeds and drop to lower levels in mature dormant seeds. Among various gibberellins, GA_3 is the most widely used gibberellin for getting rid of dormancy in seeds of many plant species. Gibberellins play a vital role in initiating enzyme activity when the quiescent embryo ingests water, later assisting in the induction of the enzyme α-amylase, which converts the embryo's starch reserves into sugars. The sugars are then transported to the growing points of the embryo, where they provide energy for continued seedling growth. Through this process, seed germination is enhanced.

b) **Abscisic acid (ABA):** It is the most common inhibitor found in dormant seeds. Its level usually increases with the maturation of fruit and is involved in the prevention of vivipary and induction of dormancy in many plant species. This substance is found in high levels in seed coat of dormant seeds of peach, walnut, apple, plum, which usually disappears when the seeds are stratified. In general, application of ABA can inhibit germination in non-dormant seeds. It is known for counteracting the effects of applied GA_3. However, these affects are temporary and disappear when seeds are placed in ABA free media.

c) **Cytokinins:** Activity of cytokinins is high in fruits and seeds during growth, but it decreases as the seed matures. It is believed that cytokinins counteract the effects of abscisic acid (ABA) during the germination process. Hence, cytokinins play a permissive role in germination by allowing GA to function appropriately. In many plant species, synthetic cytokinins can be used to break seed dormancy, such as benzyl adenine or kinetin. However, the reversal of seed dormancy by cytokinin has also been reported. This reversal in dormancy is mainly in seeds in which inhibition occurred due to natural occurring inhibitors. In such cases, cytokinin application improves germination but GA does not, for which application of both GA and cytokinin application is required to overcome seed dormancy.

d) **Ethylene:** Ethylene gas is a naturally occurring hormone that contributes to plant growth in many ways. It has been demonstrated that ethylene acts as a germination stimulant for some kind of seeds. Artificial application of ethylene has been proved to be effective in break certain kind of dormancy in seeds. Ethylene is particularly effective to overcome high temperature dormancy in seeds of certain species. It is very effective if used in combination with kinetin, CO_2 or light treatments.

e) **Other compounds:** There are certain other compounds, which may stimulate the germination process. For example, thiourea and potassium nitrate are two commonly used compounds for breaking dormancy. The role of these compounds is not clear but it is assumed that the effects of thiourea may be due to cytokinin activity. *Fusicoccin* and *cotylenin* are natural occurring compounds, which overcome seed dormancy, as does the combination of gibberellins and cytokinins.

Seed viability test

The principal objective of seed viability test is to differentiate a range of quality levels, for *e.g.*, high, medium and low vigour seeds. To test the viability, a representative sample of seed is taken from lot and one of the following methods may be used for testing the viability:

i) **Germination test:** A germination test will tell you if your seeds are still alive, and if they are they will germinate. The most commonly used method is to place seeds in blotting papers, which are taken up in petri dish or germination trays.

ii) **Tetrazolium test:** In this method, water dissolved colourless 2,3,5 - triphenyl tetrazolium chloride (popularly known as tetrazolium test) becomes red when comes in contact of living tissue. In living tissues, dehydrogenase enzymes are present which changes tetrazolium salts into red formazon. If the embryos are completely stained, it indicates full viability.

Other methods of testing seed viability are excised embryo test and X-ray tests.

Significance of seed dormancy

In plant ecology and agriculture, dormancy allows seeds to survive periods when seedling establishment would be unfavourable. A condition such as this can only be met when conditions favour survival of seedlings, as is the case with fruit plants in temperate regions. Such conditions are helpful for seed preservation in seed bank. Additionally, dormancy may synchronize the germination of seedlings to a particular time of year. Dormancy conditions can facilitate seed disposal by facilitating dormancy conditions. For example, modification of seed covering through digestive tract of a bird or other animals.

Seed invigouration

Most seeds slowly lose their germination capability as they age, due to a number of factors such as accumulation of inhibitors. These aged seeds when treated with specific chemicals like dipotassium hydrogen phosphate (KH_2PO_4), sodium dihydrogen orthophosphate (NaH_2PO_4) *etc.* at concentration of 200 ppm for 24 hours, drying to original moisture and then sowing has improved germination tremendously. Water soaking treatment also improves seed germination in papaya.

References

Adriance, G.W. 2010. Propagation of Horticultural Plants. Axis Books, Jodhpur, India, pp 71-78.

Baskin, J.M and Baskin, C.C. 2004. A classification system for seed dormancy. *Seed Sci. Res.,* 14:1–16.

Bentsink, L. and Koornneef, M. 2008. Seed Dormancy and Germination. The Arabidopsis Book, 6: e0119.

Bewley J. D. and Black M. 1994. Seeds: Physiology of Development and Germination. New York, Plenum Press.

Bewley, J.D. 1997. Seed germination and dormancy. *The Plant Cell,* 9(7): 1055-56.

Krishna, H. 2012. Physiology of Fruit Production. Stadium Press Pvt. Ltd., Delhi, India, pp 53-64.

Lang, G.A. 1987. Dormancy: A new universal terminology. *Hort. Sci.,* 22:817-820.

Manz, B., Muller K., Kucera B., Volke F. and Leubner-Metzger G. 2005. Water uptake and distribution in germinating tobacco seeds investigated *in-vivo* by nuclear magnetic resonance imaging. *Plant Physiol.,* 1386(1):1538–1551.

Roberts, E.H. 1973. Predicting the storage life of seeds. *Seed Sci. Technol.,* 1: 499-514.

4

Seed Quality, Treatment, Packaging, Storage, Certification and Testing

S. K. Acharya, Mukesh Kumar, G. S. Patel and M. K. Sharma

Seed quality is the possession of seed that is genetically pure and physically homogeneous, coupled with health and wellness.

A seed is the first genetic material that becomes the basis for future agriculture and seeds are the basic building blocks of agriculture. A great deal of farming has been modernized by technology, but without a regular and steady supply of high quality seed, yields would be drastically reduced and crop quality would decrease as well. The production of horticultural and agronomic crops depends largely on quality seed. For farmers, seed characteristics such as genetic purity, germination percentage, vigor, and appearance have a significant effect on yield. Every professional seed producer strives to achieve and maintain high seed quality.

Following are a few of the major seeds quality characteristics.

1. **Genetic purity:** The seed bears the same characteristics as its parents. *i.e.,* the seedling / plant / tree from the seed should resemble its mother in all aspects. It is important that the crop be of high quality to achieve the desired goals of increasing yield, resistance, or desired quality factors.
2. **Physical purity:** It refers to the freedom of seed from debris, diseased seeds, inert matter, and insect damage. A seed that has physical quality should be uniform in size, weight, and color. It should not contain stones, plant debris, dust, leaves, twigs, stems, flowers or any other crop seeds. Additionally, the seed ought to be free from shriveled, diseased, mottled, cracked, broken, and empty seeds. An easily identifiable seed is one that belongs to a specific species or variety. It is indirectly influenced by poor quality character when seed is planted on the field. Obtaining this quality characteristic with seed lots can be achieved by properly cleaning and grading the seed (processing) after collection and before sowing / storing.

3. **Physiological quality:** A quality of the seed that determines how it will perform in future generations. Germination and vigor of seeds make up the physiological quality characters of seeds.

   The viability of a seed refers to its ability to germinate or the time within which it can germinate. A seed's viability for producing a good seedling or its ability to produce a seedling with normal roots (radicle - the part of a plant embryo that develops into the primary root) and shoot (plumule - the rudimentary shoot or stem of an embryo plant) under favourable conditions is called germinability.

   A seed's vigour is its capacity to produce elite seedlings. The sum total of all characteristics of a seed allows it to regenerate under any given conditions. The degree of excellence of a seed or seed lot is determined by its vigour during germination and seedling emergence. It is quality seed that performs actively well in sowing and produces healthy seedlings. Based on the degree of performance in producing elite seedlings, it is classified into high, medium, and low vigor seed. The emergence rate, the uniformity of emergence, and the loss of germination are all indicators of seed vigour. Therefore, all viable seeds do not have to be germinable, but all viable seeds will also be germinable. Likewise, all vigorous seeds will germinate, but not every germinable seed will be vigorous. Physiological quality of seed can be produced through careful selection of seed (mature seed) used for sowing and by paying attention to quality characteristics during extraction, drying and storing. For raising a good plantation, it is preferable to have seeds with good vigor, since the fruits will be realized in the following years, as well as the economics (return on investment). So, choosing the right seed based on the vigor of the seed is vital for raising perfect plantations.

4. **Seed health:** A seed is considered healthy when it is uninfested with insects and free of fungal infections. A seed's health is determined by whether it has disease-causing organisms or not. Infested seeds or fungi will adversely affect the seed's physiological quality and physical characteristics while in storage. Additionally, to the health status of seed, deterioration can also be seen in the low vigor status of that seed. The health status of seed influences the seed quality characters directly and warrants their soundness in seed for the production of elite seedlings at nursery / field. It is necessary to test seed lots for health to determine whether they contain inoculum which may give rise to progressive disease development in the field and reduce the commercial value of the crop; imported seed lots may introduce new diseases. It can provide

insight into seedling evaluation and the causes of poor germination or field stand establishment.

1. Seed treatment

For most agricultural crops across the world, presowing seed treatment with systemic fungicides is a well-established practice. The treatment is designed to protect the crop against infections that are spread by seed and soil. Endophytes in seeds may help in seed development, germination, seedling establishment, and crop yield.

What is seed treatment?

In order to protect the seed from storage pests, the processed seeds are treated with fungicides and insecticides. Among the reasons for seed treatments are:

Seed disinfection: The fungicidal treatment must penetrate the living cells of the infected seed in order to kill the pathogens.

Seed disinfestations: Spores or other pathogenic organisms often contaminate seeds on their surface without penetrating or contaminating them (*i.e.*, destruction of surface borne organism)

Seed protections: To protect the seed and young plants from pathogens in storage and in the soil.

A. Seed treatment with water: The effect of water soaking participates in a number of biochemical reactions and serves as a medium for the life process. During the germination process, water is a vital component of the external environment. By soaking the seeds in water at room temperature, seed coats are softened, inhibitors are removed, germination time is reduced, and germination percentage is increased.

B. Seed treatment with chemicals: (fungicides, insecticides and botanicals)

Seed protectants

A formulation of chemicals with residual toxic or repellent properties, or both, can be applied directly to grains. Until 1950, clay minerals (red earth and kaolin) were applied directly to the grain to prevent damage from stored product pests. They are now only used on seeds and grains intended for animal feed. Grain protectants can be sprayed/dusted on uninfested grain to prevent infestation.

Halogenation seed treatment

Most seeds are treated with it during the production cycle to protect them. As part of this process, chlorine- or iodine-based halogen mixtures are prepared and applied dry or as a slurry.

Preparation of chlorine based halogen mixture

Calcium carbonate ($CaCO_3$) is dried in an oven at 100°C for 24 hours. The bleaching powder (calcium oxychloride) and *Albizia amara* leaf powder is mixed with the $CaCO_3$, in the ratio of 5:4:1 ($CaCO_3$: $CaOCl_2$: *Albizia amara* leaf powder). The mixture is kept in an airtight container for impregnation for 3-4 days. After that, the mixture is used for treating the seed @ 3 g/kg of seeds. Slurry treatment involves mixing the halogen mixture with the seed with five milliliters of water per kilogram of seed.

Iodine based halogen mixture

An airtight container is filled with three miligrams of iodine crystals impregnated in three grams of dry $CaCO_3$. Seeds are treated with the iodinated mixture at 3 g/kg.

Advantages

1. Significant improvements were made in growth and vigor.
2. Seeds should be protected from storage pests.
3. Fungi in storage are reduced significantly.
4. Enhance the field establishment process.
5. Plant growth at an early stage.
6. Chlorophyll content and leaf area increased.
7. Increase crop yield.
8. Chemicals are environmentally friendly and cost effective.
9. In this treatment, seed is converted into grain and does not have to be disposed of.
10. This enzyme reduces free radical formation, preventing natural deterioration.
11. Humans and animals are less likely to be harmed by it.
12. During storage, maintain high germination rates.

Characteristics of good seed protectants

1. To be effective, it must be able to fight diseases.
2. Easily handled and cheap.
3. Seeds are not harmed by prolonged storage.
4. The chemical is non-corrosive to machines and non-injurious to users.
5. The seed or the soil is stable in the package.
6. Inoculants are compatible with this product.

Recommendation of seed treatment for papaya

The seed treatment with copper oxychloride resulted in 63.65 % seed germination along with 96.00 % control of seedling damping off disease, followed by metalaxyl + mancozeb with 59.13 % seed germination and 94.00 % control of seedling damping off disease in papaya.

The seed quality parameters of papaya (*Carica papaya* L.) cv. Surya can be improved by extracting seeds from complete yellow/orange fruits, using GA_3, KNO_3, and *Azotobacter chrococcum* treatments.

C. Gaseous treatment

Gaseous treatment is recommended when large quantities of seeds are available for treatment. Fumigation is a treatment that uses gas under airtight conditions. The advantages are (i) it is very effective against insects and (ii) it is highly permeable. Common fumigants include:

Solid form – Celphos, quickphos,

Liquid form – CTC – LB, ethylene dibromide.

Fumigation has varying effects depending on

1. Seed moisture content.
2. Dosage and concentration of fumigant.
3. Treatment duration.
4. Degassing is important.
5. Fumigation (spot fumigation): This is performed in fumigation centers and in godowns.
6. Treatment frequency.

The advantages of seed protectants over fumigants

1. The treatment should be used prophylactically.
2. When stored in loose containers without fumigation, it can be effective.
3. Fumigants are less dangerous.
4. Germination should not be adversely affected.
5. A single application at harvest time is sufficient for one year.

Advantages of seed treatment

1.	Seeds are protected from seed rot and seedling blight with this product	*Pythium* and *Rhizoctonia* will rot seeds even before they emerge. Fungicide coatings can protect seeds from mechanical injury.
2.	Enhances germination	Managing seed-borne fungi.
3.	Protects against pests in storage	Pests caused 20 percent of the loss.
4.	Insect control in soil	Nematodes, maggots, and roots grubs.
5.	Nutritional additions	Pelletizing and coating seeds adds nutrients to the seeds.
6.	Ensure easy sowing	Increasing the size of fuzzy seed signs.
7.	Vaccination with biofertilizers / biocontrol agents	To control wilt disease in pulses, *Trichoderma viridae* increases nitrogen fixation.
8.	Removing dormancy factors	Acid heat treatments for removing hard seed coats.

Seed conditions that require treatment

Injured seed: Pathogen invasion.

Diseased seed: Infected during harvest and storage.

Undesirable soil conditions: Cold storage, damp soil prone to diseases like wilt and root rot.

Seed treatment equipment

1. Slurry treaters
2. Direct treaters
3. Home-made drum mixer

Precautions during seed treatment

Almost all seeds are treated with products that are harmful to humans, but they can also be harmful to seeds. It is vital to ensure that treated seed is never used as food for humans or animals. This possibility should be minimized by clearly labelling treated seeds as being hazardous if consumed. In order to prevent the temptation to use unsold treated seed for human or animal consumption, it is imperative to treat only the quantity for which sales are guaranteed. It is also important to treat seed at the correct dose rate; applying too much or too little material can be as damaging as not treating the seed at all. The high moisture content of some seeds makes them particularly vulnerable to damage when treated with concentrated liquid products.

If the seeds are also to be treated with bacterial cultures, the order in which they should be treated will be as follows:

i. Chemical treatments
ii. Insecticide and fungicide treatments
iii. Special treatments

D. Seed treatment with biofertilizers

This treatment facilitates the fixation of atmospheric nitrogen and preserves fertilizer usage.

Rhizobium leguminosarum biovar *trofolii*: Used in most fruit crops.

Azospyrillum, *Azotobactor*: Used in most fruit crops.

Carrier: Rice or sorghum gruel

E. Other invigorative seed treatments

Seed fortification

This form of seed treatment uses chemicals, vitamins, and minerals to energize the seed. During storage, tomato seeds are soaked in vitamin E dissolved in acetone.

Seed infusion

A seed susceptible to soaking injury and seeds with internal pathogens that require penetration by chemicals are soaked in alcohol dissolved nutrients such as vitamins and pesticides and can be dried quickly. Topsin and thiram are examples of infusions into vegetable seeds.

Seed colouring

Non-toxic dyes such as methyl red, bromocresol green, methylene blue and rosebengal dyes can be used to dye seeds. As a result, they are useful for identifying lots, varieties, brands, the A.B.R. line, and the efficiency of seed treatment. Colour faded seeds that are still vigorous but have been aged.

Seed tapes

These are tapes embedded in paper or gel. They are placed equidistantly in the gel and cooled. These tapes can be rolled out in the field, covered with soil and watered.

Seed hardening

The method is used to improve the germination of seeds sown in arid regions. Seed hardening before sowing changes the biochemical and physiological characteristics of seed protoplasm and increases the physiological activity of embryos and their associated structures. As a result, the cell wall becomes more elastic and the roots become stronger and more efficient.

Seed coating

It is a method of applying additives to the external surface of the seeds, *e.g.*, pesticide, nutrients, or nitrifying bacteria. Contrary to pelleting, coating conforms to the shape of the individual seeds and does not normally alter their size greatly.

Pelleting

Seed pelleting is a presowing management where the seed is enclosed in a filler material using an adhesive, just large enough to produce a globular unit of standard size to facilitate precision planting and to serves as a mechanism of applying needed materials on the seed.

Advantages of pelleting

1. Seed size uniformity.
2. Combining seeds and preventing clogging.
3. Precisely planning plants.
4. Growth regulators, micronutrients, and needy substances.
5. Fertilizer / nutrients localization.
6. Provide aerial seeding with ballistic ability.
7. Seeds should be protected from external organisms.
8. Seed germination remedy for problematic soil.
9. Reduces seed rate.
10. Improve field establishment.
11. Yield improvement.
12. Technology with low costs.

2. Seed packaging

In seed packaging, seeds are packed in bags after they are weighed and sewn or a sample of seeds is placed in a container following their weighing or counting, and sealed hermetically.

The ideal storage facility should satisfy the following requirements:

1. Ideally, it should provide maximum protection against ground moisture, rain, insect pests, mold, rodents, birds, *etc*.
2. Among its features should be the ability to inspect, disinfect, unload, clean, and recondition.
3. Ideally, it should protect grain from excessive moisture and temperatures that are conducive to the growth of insects and moulds.
4. In addition to being economical, it should also fit the specific situation.

When selecting packaging materials, it is important to consider the following:

1. Seed type to be packed.
2. Packing materials will be stored in a storage environment.
3. Seed quantity.
4. Storage period.
5. Amount of seed value.
6. Transport of seed.
7. Packaging materials cost.

Why?

In order to prevent insects and diseases from contaminating seeds, seeds are packaged to prevent absorption of water from the air after drying.

When should seeds be packaged?

Seeds should be packaged immediately after determining their moisture content and ensuring they are within the storage range. Seeds will always show equilibrium between their moisture content and the relative humidity of the environment and therefore, if possible, seeds should be packaged into containers and hermetically sealed in the drying room or without delay on being removed from it.

How should seeds be packaged?

Seeds can be stored in a variety of containers and with specialized sealing equipment. Ideally, the storage container should be hermetically sealed and moisture proof. The following types of containers are acceptable for base and active collections: cans, bottles, and laminated aluminum foil containers. It depends on the container and equipment your gene bank is using. The general steps outlined in this section could be followed.

Moveable racks make the best use of available space and are ideal to store containers in walk-in stores. For easy locating of individual accessions, small containers or aluminum foil packets can be arranged in boxes. Coding systems by number or color can also help in precisely locating accessions.

3. Seed storage

It can be beneficial to have some means of storing highly viable, high quality seeds based on the timing of seed collection and subsequent planting programmes. The seeds must, of course, be properly dried before storing. A lower storage temperature reduces the rate of seed respiration while seed is being stored.

It depends on the viability and the situation of the seed as to whether seed storage will be short term, *i.e.*, up to one year, or long term. Seeds can be stored either at room temperature or under refrigeration.

Short term storage

Seed may be stored at room temperature if refrigeration is not available, even if daily and seasonal fluctuations occur. Although seeds can be packaged in boxes or bags, it is better to pack them in airtight containers, keep them at room temperature, maintain constant humidity, and if necessary, cover them to protect it from insects and other pests and may put the desiccating materials (calcium chloride) in small cloth bags can be placed in the containers with the seeds to avoid the fungal infection.

Insects can be kept away from the containers if they are kept off the floor and away from walls. Additionally, the containers should be stored so that air can circulate around them, keeping the seed dry and cool. High temperatures can inhibit germination.

It is more effective to store seed in airtight containers at a constant, cold temperature near zero degrees, even in the short term. Seeds kept in cold storage should not have a moisture content greater than 4 to 10 per cent. In order to maintain a check on the moisture content, periodic measures should be taken, as follows:

A small sample of seed is removed from cold storage and weighed; dried at 105°C for 16 hours and reweighed; and, the moisture content of the seed is calculated as follows:

$$\text{Per cent moisture content (\%)} = \frac{\text{Original weight} - \text{Oven dry weight}}{\text{Original weight}} \times 100$$

In a cold storage facility, refrigeration units cool an insulated and ventilated room. It may not be possible to install such a system due to financial or other reasons, in which case an alternative is necessary. For example, refrigerators of large houses can often be used for storing at least small quantities of seeds.

A storage facility can be built without energy by using layers of straw between cork or wood planks. Generally, the temperatures in such stores will fluctuate less than those outside; in general, this is good because seed will be more viable in conditions where temperatures do not fluctuate greatly. A store of this construction, which is relatively cheap to build, should be planned in such a way that refrigeration units can be added later in the future if a source of energy becomes available.

It is extremely important to keep seed stored in an environment that is controlled to prevent temperature fluctuations and moisture changes, both of which are adverse to seed storage.

Long term storage

The discussion above regarding the short term storage of seed also applies to the long term storage of seed. In addition, the longer the period of storage, the greater the importance of monitoring the environment within a storage facility.

4. Seed certification

It is a method for ensuring the physical identity and genetic purity of registered types and varieties of crops by maintaining and making available to the general public a continuous supply of high quality seeds and propagation materials. Seed certification is a legally sanctioned quality control system for seed multiplication and production with the goals of systematic increase of superior varieties, identification of new varieties and rapid increase under appropriate and widely accepted names, and provision for continuous supply of comparable material through careful maintenance.

History of seed certification in India

In 1963, NSC was formed to conduct field evaluations and certify crops. In 1966, the first Indian Seed Act was enacted, and in 1968, the Seed Rules were formulated. As a result of the Seed Act of 1966, official Seed Certification Agencies were established in the states. As a part of the Department of Agriculture, Maharashtra established an official Seed Certification Agency during 1970, whereas Karnataka established the Seed Certification Agency as an autonomous body during 1974. The Seed Act, 1966 established 22 seed certification agencies across the country. In India, seed certification is voluntary, and labelling is mandatory.

Process of seed certification

1. **Source of planting material/ origin of the propagating material:** Verifying the source of seed is the first step in a seed certification programme.
2. **Field inspection:** Evaluation of the growing crop in the field for pure varieties, isolation of seed crops to prevent outcrossing, physical admixtures, disease dissemination and also ensure crop condition with respect to the spread of specific diseases and the presence of objectionable weeds, *etc.* by the seed certification officer or seed inspector.
3. **Sample inspection:** A composite seed sample is compared to a standard sample to determine the quality of the lot and the sample.
4. **Control plot testing:** Standard samples of the variety are grown side by side with the seedlings from the control plot.
5. **Grow out test:** Comparative evaluation of the seeds by comparing them to their standard traits for determining their authenticity to specific species or varieties.

5. Seed testing

In order to minimize the risks of planting low-quality seeds, seeds need to be tested for their planting value and to achieve the following objectives:

1. The quality of the plant, *i.e.* its suitability for planting.
2. Identification of seed quality issues and their probable causes.
3. Assess the need for drying and processing and the specific procedures to be followed.
4. In order to determine the quality of seed and its labelling.
5. The purpose of this is to establish and provide a basis for the differentiation of lots in the market regarding quality and price.

Procedure of seed testing

As soon as the samples are received in the laboratory they are entered into a register and assigned laboratory test numbers and codes, which are used in further analysis.

1. **Moisture test:** It is done after the registration of seed samples.
2. **Sampling:** The sample is divided into working samples for various tests using mechanical dividers, random cups, and spoons, or modified halving methods.
3. **Minimum weight:** For purity analyses, working samples should contain at least 2500 seeds, but should not exceed 1000 g. Each species should be counted 10 times the weight for purity, but not more than 1000 g.
4. **Purity test:** Each working sample is divided into two fractions: the selected component and the rest. Weight is assigned to both fractions, with the former being calculated as a percentage of both combined. Fraction of purity test *viz.*, pure seed, other seeds, inert matter, weed seeds *etc.*

References

Arreghini, R. I. 1972. Presowing treatment of seed of *Prosopis caldenia*. Seventh World Forestry Congress, Proceedings, (Buenos Aires, Argentina, October 1972).

Ayesha, M. S., Suryanarayanan, T. S., Nataraja, K. N., Prasad, S. R. and Shaanker, R. U. 2021. Seed treatment with systemic fungicides: Time for Review. *Front. Plant Sci.*, 12:654512. doi: 10.3389/fpls.2021.654512.

Bajpai, M. R. and Totawat, K. L. 1976. Efficacy of sulfuric acid seed treatment on germination of *Prosopis spicigera. Rajasthan J. Agric. Sci.*, 2(1): 18–20.

Bazzaz, F. A. 1973. Seed germination in relation to salt concentration in three populations of *Prosopis farcta. Oecologia*, 13(1): 73–80.

Becker, R. and Grosjean, O. K. 1980. A compositional study of pods of two varieties of mesquite (*Prosopis glandulosa, P. velutina*). *J. Agric. Food Chem.*, 28(1): 22–25.

Bhardwaj, D. K., Bisht, M. S., Jain, R. K. and Sharma, G. C. 1980. Prosogerin-D, a new flavone from *Prosopis spicigera* seeds. *Phytochemistry*, 19: 1269–1270.

Bhimaya, C. P., Kaul, R. N. and Ganguli, B. N. 1965. Studies on pre-sprouted stumps of *Prosopis juliflora. Annals Arid Zone*, 4: 4–9.

Bogusch, E. R. 1950. A bibliography on mesquite. *Texas J. Sci.*, 2(4): 528–538.

Brookbank, G. 1975. Native Mesquite Trees. Cooperative Extension Service, College of Agriculture, University of Arizona Q-355, Tucson, Arizona, p.2.

Burkart, A. 1976. A monograph on the genus *Prosopis* (Mimosoideae). *J. Arnold Arbor.*, 57(3): 219–249.

Chatterji, U. N. and Mohnot, K. 1969. Ecophysiological investigations on the imbibition and germination of seeds of *Prosopis juliflora* Linn. *Symposium on Recent Advancements in Tropical Ecology*, pp. 261–268.

Dafni, A. and Negbi, M. 1978. Variability in *Prosopis farcta* in Israel: Seed germination as affected by temperature and salinity. *Israel J. Bot.*, 27: 147–159.

Dafni, A. and Negbi, M. 1980. Variability in *Prosopis farcta* in Israel: Fertility and seed production in populations from different habitats. *Acta Oecologica, Oecologia Plant.*, 1(4): 335–344.

Gupta, A. K., Choudhary, R., Kumari, K. and Solanki, I. S. 2016. Management options for control of damping off and root rot disease in papaya. *Res. Crops*, 17(4): 763-768.

Hartman H. T. and Kester, D. E. 1979. Plant Propagation: Principles and Practices. 4th ed. Prentice Hall of India, Ltd., New Delhi

Lay, P., Basvaraju, G. V., Sarika, G. and Amrutha, N. 2013. Effect of seed treatment to enhance quality of papaya (*Carica papaya* L.) cv. Surya. *Glob. J. Agric. Health Sci.*, 2(3): 221-225.

5

Propagation Methods and Graft Incompatibility

Rajkumar Jat, Jitendra Singh Shivran, Sampurna Nand Singh and Rajender Kumar

Introduction

Multiplication of plants is termed as propagation. Nowadays, propagation practice has been totally modernized. Propagation practices have advanced from field to *in-vitro*. Starting from scion stick of 15 to 20 cm length, the plants are multiplied even using cell, tissue and embryo. In modern horticulture trade, plant propagation is a soaring business. It is highly profitable, employment generative, recreative and creative business.

Several improvements have been noticed over the period of time in the basic techniques of plant propagation, using either conventional or tissue culture approaches. The propagation under protected environment by creating artificial conditions such as light, temperature and humidity under controlled irrigation through drip irrigation and fertigation has enhanced the potential of plant propagation and has made it possible to propagate the plants around the year. The advance approaches like meristem culture, shoot tip grafting and micro budding have been commercially exploited to produce large number of healthy and true to type plants in short span.

Types of propagation

(A) Sexual propagation

Multiplication of plants by seed is known as sexual reproduction, and such plants are called seedlings. Following meiotic cell division, sexual reproduction involves the reduction of chromosome numbers to half, which returns to normal following fertilization. Sexual propagation leads to variation among progenies. (It has several advantages and disadvantages which has been given in chapter 1).

(B) Asexual or vegetative propagation

In vegetative propagation plant part such as leaf, stem or root is placed in favourable environment that it develops into a new plant. The new plant being a portion of the parent plant, are similar to mother plants. The process of mitosis division is continuously occurring in the shoot tip, the root tip, and the cambium. Chromosomes divide longitudinally during mitosis to form two daughter cells. Vegetatively produced plants possess all the same characteristics as the mother plant, being genetically identical to their mother plants (Sharma, 2002). It has several advantages over sexual propagation which has been given in chapter 1.

Propagation by seed

Seeds are typically used to propagate papaya, phalsa, acid lime and jamun. Seeds are also used for the raising of citrus and mango seedlings rootstock. On a commercial scale, this method is usually used in these fruit crops because it is easy and cheap. Nucellar seedlings can be used for mango and citrus to grow true to type plants. The propagation by seeds is important for the breeding of new varieties of plants maintaining gene pools. To grow a plant from a seed, it is essential to know the viability of the seed, its storage, sowing time, factors that promote germination, and how to care for the germinated seedling.

An embryo (or seed) in quiescent state germinates upon water absorption in the absence of any internal barrier to germination. While those seeds that do not germinate even under favourable condition of germination are known as dormant seeds. Natural or chemical dormancy can occur in fruit crops. The dormancy is due to the presence of hard seed coats that obstruct the penetration of water and oxygen needed for germination in ber, guava and walnut seed. A chemical inhibitor (abscisic acid) causes dormancy in the seeds of most temperate fruits (apple, pear, peach, walnut).

Seed dormancy: To resolve seed dormancy, soften the seed coat and other coverings. Several methods are used for breaking seed dormancy of fruit crops (This topic has also been described in chapter 3).

Sowing of seeds

Seeds are planted in seed beds, in polythene bags, or *in-situ*. Tropical and subtropical fruit seeds are sown in monsoon (June-July) or in early spring (February-March). Mango seeds are usually sown between June and July and aonla, ber and guava seeds in February and March. The seeds of temperate fruit are generally available from June to October, and should be sown after the dormancy period has ended. In citrus, mango, loquat, litchi and jackfruit, the viability of the seed is much lower, so that it should be sown immediately after extraction. Citrus seeds can be found throughout the winter months in northern India. Seed germination is low because of the low temperature prevailing. Therefore, the use of allkathene tents on seedlings in December–January is beneficial as seedlings more germinated and grow more rapidly.

Seeds are usually sown 3–4 times the size of the seeds. In light soils, it should be a little deep; in heavy soils, it should be shallow. Polythene bags and earthen pots and pans are becoming common methods of storing seeds today. Papaya seeds are usually sown in polythene sacks or earthen pots. Also, mango stones are sown in polythene bags for epicotyl grafting. When plants are grown in polythene bags, care should be taken to ensure the proper growth of the root system.

Methods of asexual propagation

Multiplication of horticultural plants can be accomplished using different methods. These include cutting, layering, stooling, budding and grafting (Bose *et al.*, 1986).

Cutting: It is a method of propagating fruit plants in which part of the plant is removed from the mother plant and replanted in a suitable media in order to create a new plant that shares all characteristics with the mother plant. This method is commonly used in plants that root easily and therefore plant propagation is very fast and cheap. Some fruit plants such as grape, plum, pear, pomegranate, fig, baramasi lemon, sweet lime and hill lemon. Cuttings are broadly classified into stem cuttings, leaf cutting and root cutting.

1. Stem cutting

For propagation, a portion of the stem is taken. It is of four types *viz.*, (i) Hard wood cutting (ii) Semi-hard wood cutting (iii) Soft wood cutting and (iv) Herbaceous cutting.

Hard wood cutting: Hard wood cuttings are taken from mature trees when the tissues are fully developed. Shoots that are one year old or older are ideal for preparing hard wood cuttings. Cuttings are made after pruning (December-January) of deciduous fruit plants such as grape, pomegranate, plum, pear, phalsa, and fig. The pruned wood can be used for preparing such cuttings. In sweet lime, hill lemon and baramasi lemon, the cuttings can be prepared during the spring (February- March) and rainy season (August-September).

Cuttings are generally made with a length of 15-20 cm and 3-5 buds. A slanting lower cut is given just below the bud to increase nutrient absorption. To avoid the upper bud drying, the upper cut is made at a right angle (round) to reduce the size of the wound. After the cuttings have been prepared, they should not be allowed to dry. A callus is formed by burying 20-25 cuttings in moist soil/ sand for a period of time, causing the wound to heal. Plant growth substances such as IBA, NAA have used for root initiation in hard wood cuttings. In some plant species, rooting is very difficult which can be enhanced by treating cutting with auxins (Hartmann *et al.*, 2002).

Semi hardwood cutting: Semi-hardwood cuttings are prepared from 4 to 9 months old shoot of semi-hard nature. but tender woody shoots. Citrus and olive can be propagated in this way. The cuttings have a length of 7.5 to 15 cm with leaves at the top. Cutting is usually done at the shoot ends, but the roots are usually found at the stem's base as well. It is best to obtain cutting wood in the early morning hours when the stems are turgid. Intermittent moist sprays of water and treatment with auxins have been reported to be beneficial. This type of cutting is usually practised for evergreen plans like mango, guava, jackfruit, *etc.* Treatment of IBA @ 5000 ppm is effective in inducing rooting in such type of cuttings.

Soft wood cutting: This type of cutting is not practised for fruits. Humidity requirement being very high, practising this cutting is not possible in open condition. Soft wood cutting is practised under mist chamber. Shoots of 2-3 months age are selected for soft wood cutting. Cuttings should not exceed 10-15 cm in length. Apple, peach, guava and many ornamental plants can be propagated under mist chamber using soft wood cuttings.

Herbaceous cuttings: Herbaceous cuttings are the tender succulent special leafy part of the stems of herbaceous plants. A healthy shoot's terminal 7.5 to 12.5 cm are cut off, and the basal leaves are removed, leaving the upper leaves undisturbed. They are rooted under the same conditions as softwood cuttings, requiring high humidity. Herbaceous cuttings of some plants exhibiting sticky sap, such as pineapple, perform better if the basal ends are dried for a few hours before inserted into the rooting medium under low light conditions. The application of auxin stimulates adventitious root regeneration.

2. Root cutting

The plants capable of producing suckers are good for root cutting. It is followed in apple, pear, cherry, guava, wood apple *etc.* Cuttings should be taken from root pieces of young stock plants during late winter or early spring when roots are well-supplied with stored food, but before new growth begins. It is important with root cuttings to maintain the correct polarity when planting as the new shoot develops from the proximal end *i.e.* from the part close to the crown. The proximal end of the root piece should be up at all times. Shoot emerges from adventitious buds on root cuttings and root emerges from cambium tissues.

3. Leaf cutting

Leaf cuttings have very little application on fruit crops. The leaf is separated from mother plant and planted in suitable medium where it gives out roots and generates a complete plant.

Anatomical basis of rooting of cuttings

1. Anatomy of the part of plant from where the cutting has taken plays a vital role in the process of root induction. The roots may be developed on stem, root or leaf cuttings but they all have internal origin in their parent structures. The adventitious roots developed from a cutting are of two types depending on the genetic makeup of the mother plants. In few plant species, pre-formed roots develop naturally on stem while they are still attached to the parent plants called as pre-formed roots. While, in certain cases roots develop only after the cutting is made, in response to the wounding effect in preparing the cutting called as wound roots. Thus, internal origin of roots is called as endogenous and was reported in most of the plant species.

2. However, in some species (*e.g.*, Tamarix), roots may first develop exogenously on the stem and then this are connected to the internal tissues of the stem. First study on anatomical basis of rooting was done by a French dendrologist, Duhamel du Monceau in 1758 and later various research workers emphasized that in the process of rooting of stem cuttings, generally four anatomical changes are observed. These anatomical changes are. i) dedifferentiated specific mature cells, ii) formation of root initials from the dedifferentiated specific mature cells, iii) development of root initials into organized root primordia and formation of vascular connection between root primordia and conducting tissues of the cuttings, emerging through the cortex and epidermis and iv) emergence of roots outside the cuttings. However, herbaceous cuttings have different anatomical changes during the rooting process, because of entirely different structure of the stem. The herbaceous stem generally has four major areas *viz.*, a large pith in the centre, a ring of vascular bundles outside the pith, a cortex outside the vascular bundles and a thin epidermis.

3. In such plants, adventitious roots generally appear just outside and between the vascular bundles but the tissues involved at the site of origin

of roots vary widely, depending upon the kind of plant. For example, adventitious roots in tomato and pumpkin arise in phloem parenchyma region, from epidermis in Crassula, from pericycle in coleus and in castor these arise from vascular bundles. In woody plants, one or more layers of xylem and phloem are present and adventitious roots are formed in the stem cuttings from the living parenchyma cells. Generally, the origin and development of adventitious roots takes place next to and just outside the central core of vascular tissues. After emergence, the roots develop root cap and other tissues of the root. Adventitious root and shoot usually arise within the stem (endogenously) near vascular tissue, outside the cambium.

Physiological basis of rooting of cuttings

Any phenomenon occurs in the living system is scientifically considered as a resultant of genotype of the organism and its interactions with the existing environment which creates specific physiological conditions within the organism to perform specific function. Although, the process of rooting is governed by genetically and environmental factors but understanding about the real physiological processes is very important.

Several physiological processes occur during the rooting of cutting

1. **Growth regulators:** Various classes of growth regulators, such as, auxins, cytokinins, gibberellins, abscisic acid and ethylene influence rooting of cuttings. However, of these, auxins are known to have greatest effect on root formation in the cuttings. In addition, various other natural occurring promoters and inhibitors may also take part in the root initiation process.

2. **Role of vitamins:** Endogenous optimum level or exogenous application of vitamin promotes root initiation process in the cuttings. Thiamine (vit B1), pyridoxine (vit B6), niacin and biotin B-complex vitamins and vitamins K or H, all are known to stimulate the rooting process in different plants. Vitamin, if applied with some auxins like IAA, NAA or IBA have synergistic effect on root initiation in cuttings of different plant-species.

3. **Presence of buds and leaves:** It has been demonstrated by many workers that presence of leaves on the cuttings exerts a strong stimulating influence on the root initiation process in cuttings. The leaves produce carbohydrates, which are translocated downward to the base of stem, where it promotes root formation. Further, the leaves and buds are powerful auxin producers and their transport to basal parts may initiate

rooting faster as in poplar, grape and currant. If the buds are removed from the cuttings, the root formation may be hindered adversely. Rooting co-factors: Many rooting co-factors have been isolated from cuttings of different plants. These co-factors are naturally occurring substances that appear to act synergistically with auxin, primarily IAA, for root formation in cuttings. Bouillenne and Went were the first to name the root- forming factors of leaf, cotyledon and buds as rhizocaline in 1933.

4. **Nutritional factors:** The nutrient status of plant from which cuttings are taken, also plays vital role in root initiation process of the cuttings. Low nitrogen and high carbohydrate resources usually favour root initiation process compared to high nitrogen and low carbohydrate reserves. High nitrogen content usually favours luxuriant growth and hinder root initiation process but extreme N-deficiency also hinders the rooting process. Thus, high C: N ratio should be maintained in the stock plants before taking cuttings from them.

5. **Endogenous rooting inhibitors:** Additions to root promoting substances, certain endogenous root inhibiting substances are also present in some plant species, inhibiting the process of root initiation. Therefore, cuttings of certain plants do not root easily because of presence of endogenous inhibitors. For example, the cuttings of *Vitis berlandieri* do not root easily as compared to the cuttings of *Vitis vinifera* because of presence of higher concentration of abscisic acid. Similarly, hardwood cuttings of Bartlett pear do not root easily as compared to Old Home cultivar. Placing these cuttings in water before planting help in leaching of inhibitors and enhance the root initiation and subsequent development of roots.

Budding: Bud grafting refers to inserting only one bud into a rootstock. In budding, the bud wood is greatly reduced as compared to grafting. Budding is done for citrus, peach, ber, almond, pear, and plum. It is usually done during the spring and rainy season. As a result of continued division of cambial cells and increased chances of bud union with rootstock, taking out bud from scion stick becomes easier during this period. This shows that the cambium which is the tissue responsible for union is active. In general, rootstocks of 1 to 2 years of age with pencil thickness are used for budding.

Types of budding

1. 'T' – budding

This is also known as shield budding. In this method, boat shaped bud of 2.5 to 3.0 cm length is used for budding. If the bud is inserted by making vertical incision on rootstock, it is termed as shield budding. If T-shape incision is made for inserting bud on rootstock, it is termed as T - budding.

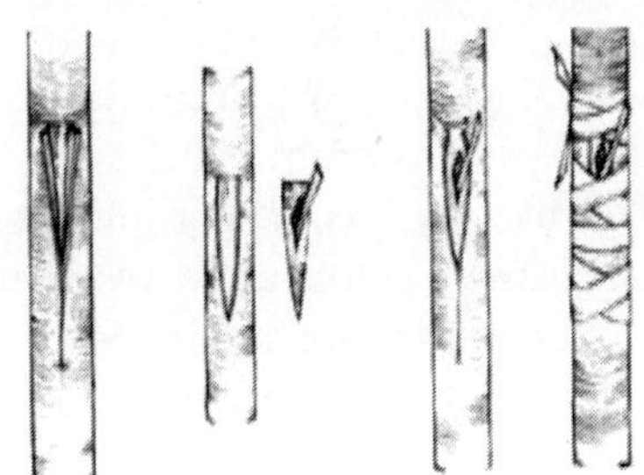

The budding is made at 10 to 25 cm height on rootstock. The bud is inserted into an incision on rootstock, then wrapped air-tight with polythene tape, leaving the bud exposed. This is the most common method of budding used in most of the commercial fruits like citrus, apple, pear, peach, almond, ber, aonla *etc.* If cell-sap flows freely, T-budding can be performed at any time of the year. In most fruit trees it is performed either in the spring (March-April) or in the rainy season (July-September) period. In *in-situ* orchard establishment of ber, I-budding was found successful. However, July to September months was suitable for budding in the field under arid conditions (Nath *et al.*, 2000).

2. Patch budding

This method is very successful for propagating plants having comparatively thick bark. Fruit plants like jackfruit, aonla, mango, jamun, chestnut *etc.* are propagated by patch budding. From a scion shoot, a rectangular or square-shaped bud is removed. On the rootstock, a similar incision is made. The bud is then attached to the rootstock.

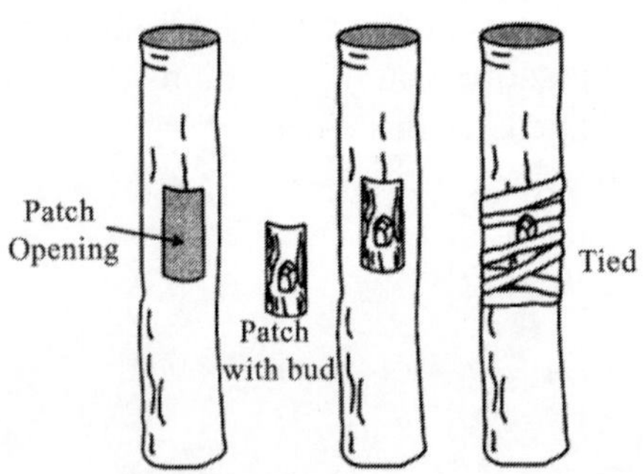

An incision of 2-3 cm is made for the placement of the bud. The sprouting portion of the buds is wrapped in polythene tape. Wrapping prevents the buds from desiccating and promotes sprouting. Aonla is commercially propagated through patch budding from June to September.

3. Ring budding

Ring budding is similar to patch budding, the only difference being that cylindrical portion of bark about 1.5 cm in thickness is removed instead of a rectangular or square patch. The ring of bark containing the scion bud is placed smoothly into the cylindrical slot by removing the bark from the rootstock. A plastic tape is then applied to the bud wood.

In ring budding, when the scion bud fails to grow the top portion of the rootstock also dies because it is girdled. This method is followed when bark slips freely. This method is commonly used for budding peaches, almond, ber and mulberry.

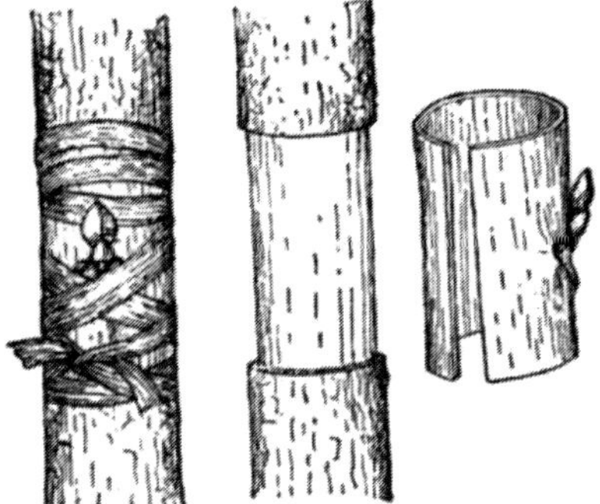

4. Chip budding

This method is usually used when the stock and scion are still dormant, just before they begin to grow. Against Phylloxera insect, grape varieties are propagated during dormant season using chip budding. During February-March apple, pear and grapes can be propagated through this technique. The bud is taken out from scion shoot along with wood. The rootstock is incised in a similar manner. A polythene tape is used to wrap the bud around the rootstock.

5. Forkert budding

In this, a transverse cut about 1.25 cm long is given on the rootstock at an appropriate height. Two vertical cuts, one from each end, are made to a length of about 1.5 cm. The bark so demarcated is carefully peeled out but it remains attached with the rootstock on the lower side. The scion-bud of a size corresponding to the size of cut made on the rootstock is removed and fitted into the exposed part of the rootstock. This bud is covered with the bark flap of the rootstock and tied with polythene strip. After about 4 weeks when the union between the bud and the stock has occurred, polythene wrapping is removed and flap is also cut off. As the bud starts growing, rootstock top is also cut off. This method is followed in propagation of mango and aonla plants.

6. Micro budding

This technique involves budding young rootstock of about 3-4 month old with small scion buds *in-vivo*. A unique way to keep micro-bud on place involves the use of micro tips/ plastic tips at initial stages after budding and later on removing them on seeing the scions of visible bud growth. This method has several advantages over *in-vitro* grafting, namely easy in performing budding operations, better success rate and practically more number of plants can be produced. However, if the scion is not taken from pathogen free plant, the resultant plant can also be infected, while in case of meristem culture, having the size of meristem 0.1-0.5 mm is in most of instance free of viruses. The basic principle is to understand that even plants raised through meristem culture are free of viruses but are not virus resistant.

Grafting: Grafting is a propagation method in which the scion stick (shoot containing more than one bud) and the rootstock are joined in such a way that they can combine and then grow and develop as a successful plant. Grafting is usually adopted in mango, guava, pear, peach, plum, almond, loquat, sapota, *etc*. Grafting is performed in peach, plum and almond at dormant stage but in mango, guava, loquat and sapota, the operation is carried out at when active growth stage of tree. There are some methods of grafting commercially followed in fruit plants.

Attached or approach method of grafting

1. Inarching

This is called as attached method of grafting. In contrast to other method, in this method the scion is detached after completion of union. This method is practiced in mango, jackfruit, sapota, loquat and custard apple. In this method, the rootstock is grown in pot or container.

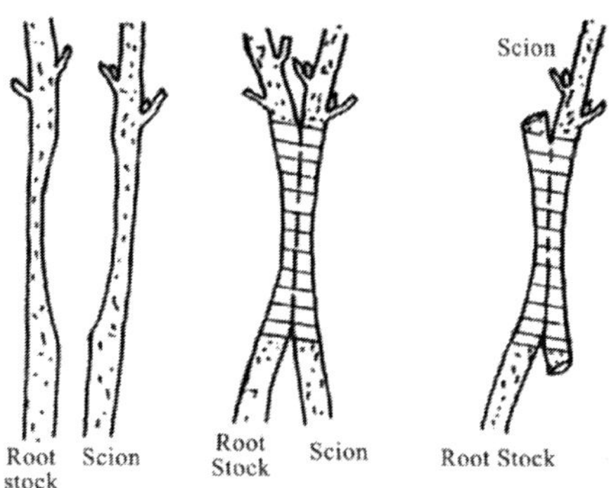

In this process, the rootstock is brought closer to the scion. The rootstock and scion shoots of pencil thickness are selected. Shallow cut of 2.5 to 4.0 cm length is prepared on scion. Similar matching cut is also given on rootstock. Both rootstock and scion are brought close to each other and wrapped using polythene tape. When union takes place, the upper portion of rootstock is cut. Scions are separated from the mother plant. The united portion is kept in nursery and cared till next planting season. The process of inarching is practised during July-August during which high humidity favours better union.

2. Tongue grafting

Grafting nursery stock is very useful with this method. Top working with tongue can also be done when the plants to be top-worked are quite young with branches only one centimetre in thickness. To achieve this, stocks and scions should have equal diameters. First a long, smooth slanting cut of about 4 to 5 cm long is made on the rootstock but if the top growth is quite heavy then it is better to remove the top with the secateur before making the slanting cut. There is another downward cut of about a centimetre in length that begins approximately 1/3 from the top. The scion wood is cut exactly the same way as the rootstock. The scion with 2 or 3 buds is then tightly attached to the rootstock, so that the cambium layer on at least one side of the stock and scion are joined. This is then wrapped with a strip of polythene. This method is used for propagating apple, pear and walnut.

Detached method of grafting

1. Veneer grafting

This method is used for propagating mango. In this method, a terminal shoot of 10-15 cm length having pencil thickness is used as a scion. The swollen shoot is used as scion. About 10 days before grafting, the scion-shoot is defoliated to facilitate swelling of bud. Shallow, downward and inward cut ensuring V-shape incision in lower portion of rootstock is prepared. Similar matching cut in slanting manner is prepared on lower portion of scion. Both the rootstock and the scion are tied with polythene strips. During May and September this system is quite successful and good result is obtained. In about 3-4 weeks, union is completed. When scion sprouts completely, the upper portion of rootstock above graft union is removed in 2-3 instalments.

2. Whip or slice grafting

This is very simple technique of grafting. Generally, one year rootstocks are used and scions of matching thickness are used. Polythene tape is used to bind rootstock and scion together. When union is completed, the polythene tape is cut otherwise it restricts the growth at the point of union and such plants break due to wind. It is used for propagating walnut, apple and pear.

3. Cleft or wedge grafting

In nurseries, where the rootstock is thicker than the scion and tongue grafting cannot be successfully used, this method is useful. Rootstock with 5–7 cm or more girth, should be selected and cleft grafted after decapitating the stock 45 cm above the ground level. The beheaded rootstock is cut 5 cm deep through the centre of the stem with a knife and a mallet. A wooden wedge is inserted after the knife is removed to keep the hole open for the subsequent insertion of the scion. Scion (15–20 cm) is used from a terminal shoot which is more than 3 months old and it is wedged securely (6–7 cm). A wooded wedge is used to force the split in the stock to receive the cleft of the scion. Walnut, hazelnut, peanut and grape are propagated by this method. In addition, this method is commonly employed for rejuvenating the old temperate orchards by top working.

4. Epicotyl or stone grafting

The method is commonly practised in mango. A layer of leaf mould covering 5-7 cm of the stones is sown in a moist sand bed for germination. We take the seedlings out of the ground and graft them indoors after 15 days. In order to remove the top of the seedling, a slanting cut at a height of 5 cm is made. The scion shoot of 8-10 cm long from current season's growth is prepared by giving a slanting cut at the base in one side so as to match with the cut end of stock. It is then placed on the stock and tied with polythene tape. Following grafting, the seedlings are planted in polythene bags containing soil, sand, and farmyard manure in a 1:1:1 ratio, and watered immediately. The polythene bags are kept in a partial shade condition and watered daily. When the scion produces four true leaves, they are transplanted in the nursery bed.

5. Soft wood grafting

In-situ grafting is a very successful technique. It is commonly practised in mango. During the rainy season, the mango seeds are sown in the field at the desired distance. To ensure germination, 2 to 3 seeds are sown in each pit. After the plant has reached pencil thickness and is one year old, it can be grafted. In the field, the grafting is done at the permanent planting site. Rootstocks are grafted during the rainy season when new growth appears. Grafting is necessary when new growth begins to turn yellow from coppery colour. Scion shoot of 10 to 15 cm length; 3 to 5 months of age and pencil thickness girth is selected. At 15 to 20 cm height from ground level, the rootstock is beheaded. On rootstock, a vertical slit is made between 2.5 and 4.0 cm long. In the lower portion of the scion shoot, a similar matching cut is prepared in a slanting manner on both surfaces. A polythene tape is used to wrap it around the incision on the rootstock. In about 3 to 4 weeks, sprouting starts and graft starts growing. The grafted plant develops at its own root system and shows better survival in the field.

6. Bridge grafting

This method is practised in plants in which scion is healthy and some portion of rootstock near collar region is damaged. In this technique, the damaged portion of rootstock is scratched. In healthy portion of rootstock incision is made on top and bottom portion of the stock. The scion portion of suitable length is inserted into incision. Nails are used to fix it, and wax is used to seal it. The sprouted bud from inserted stick should be removed time to time. Slowly and slowly, it grows in diameter and cover the damaged portion. This method is useful in repairing damaged wood in apple, pear, cherry, walnut *etc.*

Layering: Layering is a plant propagation method, easy to use for the multiplication of fruit plants, in particular those that do not easily root from cuttings or grafting. During layering, roots are induced on the shoots while they are still attached to the mother plant. In this process, the root and shoot form a single portion of the plant. When the roots develop, the shoot is separated from the mother plant and has survived for some time in the nursery and is then planted in the field.

Types of layering

1. Simple or ground layering

In this method, the ground touching portion of plant and suckers emerging along with main stem of the plants are used for layering. Generally, one year old shoot is used for layering. The shoot is bent downward in the soil using peg or nail and tied with the help of rope to make it stayed in the position. The ground touching portion is wounded. In about 4 to 5 months, roots emerge out. It is then separated from mother plant by giving 2 or 3 cuts in instalment. Guava, lemon and limes can be propagated easily through this technique.

2. Serpentine or compound layering

Covering the branches of plants at their nodes with soil throughout its length by alternate exposing of internodal length of shoot is termed as compound layering. It is a modification of simple layering in which a branch of one year is alternately covered and exposed along its length. The underground part is girded at various points. To develop new shoots, at least one bud should be present on the exposed portion of the stem. Once the sections have rooted, they are cut and planted in the field. It is used for propagating muscadine grapes.

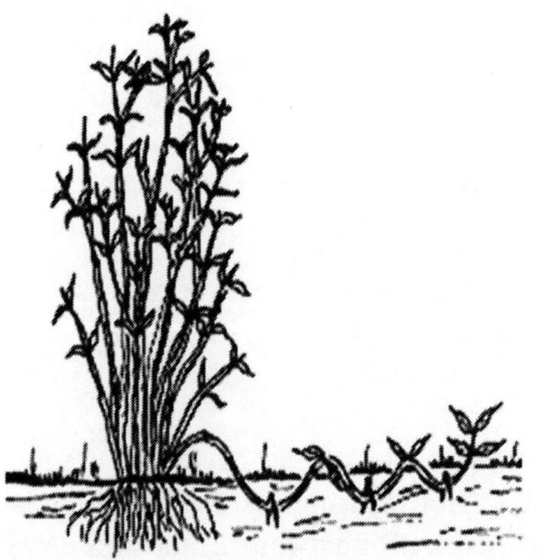

3. Mound layering or stooling

During the dormant season, the plant is moved back to a height of 15 cm above the ground. Within two months, new sprouts will appear. Sprouts are then girdled near the base, and rooting hormone (IBA), derived from lanolin paste, is applied to the upper portion of the cut with moist soil. Before they are covered with moist soil, these shoots are left as such for two days so the rooting hormone can be properly absorbed. Generally, rooting hormone concentrations range between 3,000-5,000 ppm, depending on the plant. Rooting typically occurs between 20 and 30 days after transplant. After 2 months, the rooted stools are separated from mother plants and planted in nursery. Guava and apple are commercially propagated in this way. In addition to plums, cherries, hazelnuts, pecans, mangoes, jackfruit, and litchi, this method is also advocated in other fruits.

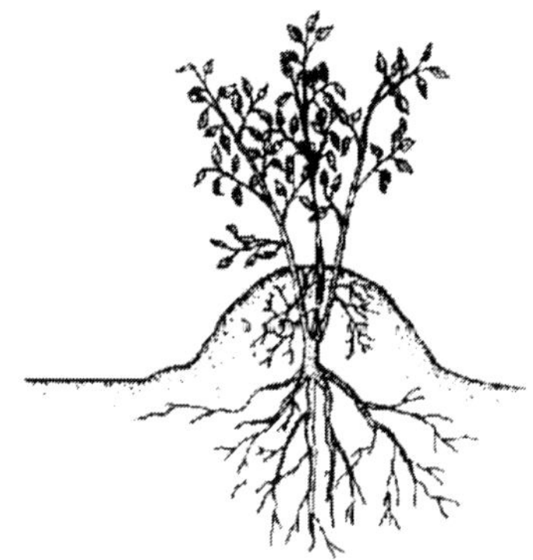

4. Tip layering

It is generally followed in plants which have trailing type of shoots. In this type of layering, tips are buried to a depth of about 7.5 cm or they may be buried in pots filled with rooting medium leaving a few cm of the shoot tip above ground. The covered portion strikes roots within 2-3 weeks. The rooted layer is then removed and transplanted either in the nursery bed or in a pot. This method of propagation is used usually in black berries, raspberries, and gooseberries.

5. Air layering

It is also known as Chinese layering, Pot layering, Marcottage or Gootee. One year old or previous season shoots of pencil thickness are selected for air layering. About 5 to 7 cm away from the base of selected shoot, a girdle of 2.5 to 3.0 cm size, by removing the bark, is prepared. The girdled portion of the shoot is scrapped using gunny bag or with rear side of the blade. During this process, phloem is removed and ultimately bark is prevented from forming on the girdled portion. The girdled portion is then covered using moist sphagnum moss grass. The girdled portion is now wrapped using transparent polythene tape and both the ends of tape are tied air-tightly.

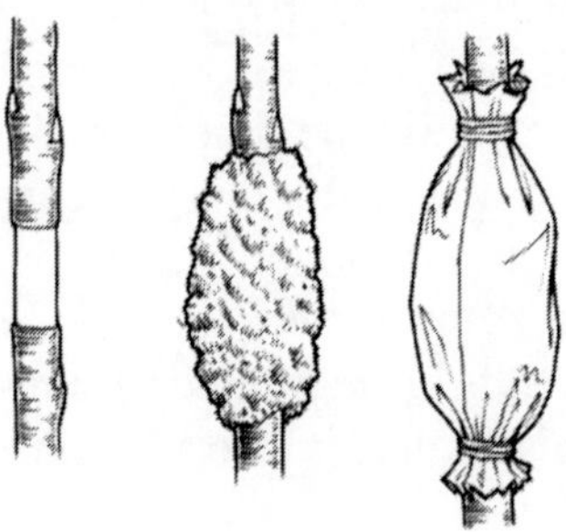

Depending upon ease of success, rooting appears in about 2-3 months. When root emergence is there and it becomes visible from the transparent wrapped tape, the layered shoot is separated from mother plant giving 2-3 cuts in instalments. The layered plants are planted in nursery under partial shade where frequent watering is provided and high humidity is maintained. In guava, litchi, lime, sapota, loquat, *etc*., air layering is usually done during Feb-March and July-August.

6. Trench layering

It is used for the propagation of apple, pear and cherry. In this method, the branch of the plant is bent downward in a horizontal position in a shallow trench. When the shoot develops, the base is covered with 5 to 10 cm layer of soil. Covering the shoot with soil gives etiolation effect and helps in rooting. The rooted layers are separated from the mother plant and planted in nurseries. New roots form at the base of new shoots. Vigorous rootstock of apple like M16 and M25 and walnut are usually propagated by trench layering.

Table 1: Commercial method of propagation in different fruit crops

Fruit crop	Commercial method of propagation	Time of propagation
Almond	Shield budding	July – Aug
Aonla	Patch budding	June – Sep
Apple	Tongue grafting	Feb – March
Apricot	Tongue grafting	February
Bael	Patch budding	June – July
Banana	Suckers (sword)	March
Ber	T – budding	June – Sep

Cherry	Tongue grafting	Feb – March
Citrus	T – budding	Feb – March, Aug – Sep
Date palm	Offset (sucker)	Feb – March
Fig	Hardwood cuttings	Jan – Feb
Grapes	Hardwood cuttings	Jan – Feb
Guava	Wedge grafting	July – Sep
Litchi	Air layering	July – Sep
Loquat	Inarching	July – Aug
Mango	Veneer grafting	July – Sep
Peach, Pear, Plum	Tongue grafting	Dec – Jan
Pomegranate	Hardwood cuttings	Jan – Feb
Sapota	Inarching	Dec – Jan
Strawberry	Runners	June – Sep
Walnut	Patch budding	July – Aug

(*Source*: Chadha, K. L., 2019)

Propagation through specialized organs

Division: It is a method of asexual propagation by which vegetative part that separate naturally from the parent plant is used. These include rhizomes, suckers, offsets, crowns, runners, *etc.*

1. **Runners:** Runners are branches that grow from the axil of a leaf on a plant's crown and form new plants at one node, such as strawberry plants. Long days and high temperatures favour runner production. New plants are grown from daughter plants.

2. **Suckers:** A shoot arising on an old stem or underground part of the stem or on horizontal root systems is known as sucker *e.g.*, datepalm, pear, banana and pineapple. Separated from the mother plant, these shoots produce adventitious roots when transplanted. Banana sword suckers are commonly used to propagate plants.

3. **Offsets:** In certain plants, such as pineapples and date palms, an offset is a lateral shoot or branch that extends from the base of the main stem. With a sharp knife, offsets are removed from the main stem.

4. **Crowns:** An epidermis is a very short part of the plant on the surface of the ground that gives rise to new shoots. The crown is the general transition zone between stems and roots, located near the ground surface. Crown grown on top of the pineapple fruit is a vegetative growth attached to the central core of the fruit which is used longer duration to bear fruits

as planting material. Splitted crown into four to eight pieces, can also be used but they take longer duration to bear fruits.

5. **Slips:** Slips are shoots that appear above the ground just below the crown. In general, pineapple propagates by means of slips.

6. **Rhizomes:** A rhizome is an elongated, horizontal fleshy underground stem having nodes and internodes bearing buds in the axils of reduced scale leaves *e.g.*, banana. The rhizomes are cut into pieces, each with one or two viable bud and planted in favourable atmospheric conditions. Adventitious roots and shoots develop from the nodes.

Rootstock for fruit crops

Most of the commercial fruit crops are grown commercially as grafted plants. In other words, most of the commercial fruit crops are grafted onto rootstock (Table 2). Plants to be used as rootstock may belong to same or different species

Table 2: Most commonly used rootstock for commercial fruit crops

Fruit crop	**Rootstock**
Almond	Almond, peach seedlings
Aonla	Desi aonla seedlings
Apple	Crab apple seedlings, clonal rootstocks
Apricot	Wild apricot, wild peach seedlings
Bael	Bael seedlings
Ber	Desi ber / Malahber
Cherry	Mahaleb seedlings
Citrus	Rough lemon, *Jatti khatti*
Grapes	Dogridge
Guava	L-49
Mango	Seedlings of desi mango
Peach	Wild peach
Pear	Kainth
Plum	Kabul Green Gauge
Sapota	Rayan
Walnut	Walnut seedlings

(*Source*: Chadha, K. L., 2019)

Graft incompatibility in fruit crops

Graft incompatibility refers to the failure of rootstock and scion to produce a successful graft union. If the graft union is successful, it is known as graft compatibility. There is no clear-cut difference between a compatible and an incompatible graft union. Two plants can unite successfully into a graft because of their natural relationship. Plants of closely related species unite easily and grow as a composite plant, whereas plants of unrelated species do not. There are instances when stock and scion of unrelated species mate, but later develop symptoms of incompatibility and die. The majority of graft combinations fall in between these two extremes in that they unite initially but gradually develop signs of incompatibility such as abnormal growth patterns. In other words, graft union success differs greatly among plants. Symptoms of incompatibility may appear shortly after grafting or may take time to develop.

Types of graft incompatibility

Graft incompatibility is of two types 1. Localized incompatibility and 2. Translocated incompatibility.

1. Localized incompatibility

Incompatibility reactions of this type are believed to be triggered by the actual contact between the stock and the scion and are localized in their nature. By inserting an interstock between the stock and scion, which separates the incompatible components of the graft, this type of incompatibility can be overcome. A weak union structure occurs when stock and scion have incompatible union structures, resulting in discontinuity in cambium and vascular tissues. There may be poor translocation of metabolites across the graft union as a result. According to the degree of anatomical differences at the site of graft union, external symptoms appear slowly. Usually, masses of undifferentiated parenchymatous tissues are commonly found at the graft union this type of incompatibility or even inclusion of bark tissue may develop, which may disrupt the normal vascular connection between the stock and scion. When apples are grafted on pear trees and plums on cherries, localized incompatibility reactions are quite common. Bartlett pear grafted onto quince rootstock is an example of this type of incompatible reaction. When 'Old Home' pear interstock is inserted, the combination becomes compatible, resulting in successful tree growth.

2. Translocated incompatibility

This incompatibility reaction is sometimes caused by a labile influence that moves throughout the graft union. The insertion of a mutually compatible interstock cannot correct this type of incompatibility reaction. The bark of this type usually shows a brown line and necrotic area caused by phloem degeneration. Carbohydrates are restricted from moving through the graft union, with an accumulation above and a reduction below.

One of the most important examples of this category of compatibility is the Hale's Early peach grafted onto Myrobalan B plum rootstock. A weak union forms as a result of starch accumulation at the base of peach causing distortion of tissues. If a mutually compatible interstock, Brompton plum, is inserted between them, the incompatibility remains, with starch accumulating at the base of the Brompton interstock. When the same combinations are evaluated at the cotyledonary stage, they are highly compatible, suggesting that some of the factors responsible for incompatibility reactions are absent at the juvenile or nursery stage of the plants. The combination of Non Pareil almond and Mariana 2624 plum exhibits complete phloem breakdown, while the xylem tissues are normal. The Texas almond is highly compatible with Mariana 2624. When a piece of Texas almond is inserted between Non Pareil and Mariana 2624, bark disintegration results, resulting in an incompatible graft union. A third type of virus induced incompatibility is also apparent and widespread and more and more cases are being discovered. It is possible for pathogens to cause failed graft unions. In certain parts of the world, sweet oranges budded on sour oranges, while in other regions they were successful. According to thorough studies, incompatibility was primarily caused by viruses. In addition, compatibility reactions change over time. In 1932, the pear cultivar Bristol Cross was grafted on Quince but 30 years later it was required to graft on Beurre Hardy to form an acceptable union. Mutations or latent viruses can cause such changes in plants. During some cases of delayed incompatibility symptoms, the initial graft union was successful, but later they showed signs of incompatibility, and typically a clean breakage was observed at the point of union. For example, black line in walnut.

Causes of graft incompatibility

Despite the fact that grafting is purely a physical phenomenon, compatibility and incompatibility reactions are determined by genetic differences between the stock and scion. The mechanism for expressing a particular case, however, is unknown. It has been proposed or attempted to explain the causes of incompatibility, but the evidence supporting them is inadequate and inconsistent. Incompatible reactions can be caused by structural, physiological, biochemical, disease, insect, or a combination of these factors.

Structural or anatomical reasons*:* In histological studies, it has been found that stock and scion may not differ structurally but abnormalities may be found due to the presence or development of parenchymatous cells at the graft union, preventing the formation of vascular continuity between the stock and scion. Graft failure may occur when a bark layer develops at the joint. Distortion of vascular tissues between stock and scion also takes place due to the development of some whorls or loops, which restrict the movement of essential nutrients and water across the graft union, resulting in poor growth or failure of a graft union. Similarly, from micro-spectrographic examination of the cell walls of incompatible graft union, it was found that the cell wall adjoining cells are lacking in lignin, whereas the cells in the compatible forms were lignified. Therefore, any reaction inhibiting the formation of lignin and the development of a middle lamella between the stock and scion results in week union. Peaches, pears, and plums are primarily incompatible due to their structural differences. Usually, the development of undifferentiated tissues at the graft union leads to mechanical weakness, resulting in disintegration of the phloem and uneven or poor growth of the graft.

Physiological and biochemical reasons: In certain instances, the supply of necessary components is insufficient, either from stock or scion. As a result of incompatible unions, water supplies are often compromised. A scion may die due to root starvation caused by failure of the phloem or xylem at the union. Sugars are translocated from scion to stock in compatible unions. Assimilates tend to accumulate in the scions of incompatible combinations. A cyanogenic compound prunasin found in quince (not in pear) is translocated to pear phloem when certain pear cultivars are grafted onto quince. One of the products of the breakdown of the pear tissues is hydrocyanic acid. In the phloem and xylem of the graft union, this acid accumulates, causing tissue breakdown and decreasing cambial activity. The result is that water and other materials cannot be transported effectively. Therefore, toxic chemicals can inhibit the growth of other organisms or kill them. There is no doubt that dwarfing effects of rootstock on scion cultivars are due primarily to the restricted supply of water and nutrients. When budded or grafted onto a dwarfing rootstock, an outgrowth is commonly observed above or at the graft union. As a result of this outgrowth, nutrient and water supply to the top of the plant is restricted, resulting in dwarf growth.

Nutritional deficiency: It is also possible that incompatible unions are caused by nutritional deficiencies. The Jonathan apple has been found to develop molybdenum deficiency when grafted onto EM-IX rootstock. This may be a result of the EM-IX rootstock's inability to absorb Mo in sufficient quantities to supply it to the scion cultivar. In contrast, Jonathan's deficiency symptoms do not appear on other rootstocks. There has also been reported deficiency of P, K and Mg in peaches when grafted onto Myrobalan B plum rootstock, resulting in incompatibility symptoms.

Presence of viruses: Graft unions may fail if latent viruses or mycoplasma like pathogens are present. Due to the presence of viruses at the graft union, Bartlett trees grafted on *Pyrus pyrifolia* rootstock are prone to pear decline disease. However, it does not appear that *Pyrus communis* is used as a rootstock. Further, the incompatibility reactions in citrus are primarily due to infection by viral diseases like tristeza, psorosis and xyloporosis *etc.* In Punjab, the failure of rough lemon rootstock for many sweet orange varieties has been attributed to the occurrence of bud union increase viral disease. The delayed incompatibility reaction in black line walnuts is also believed to be the result of a virus and not a rootstock failure.

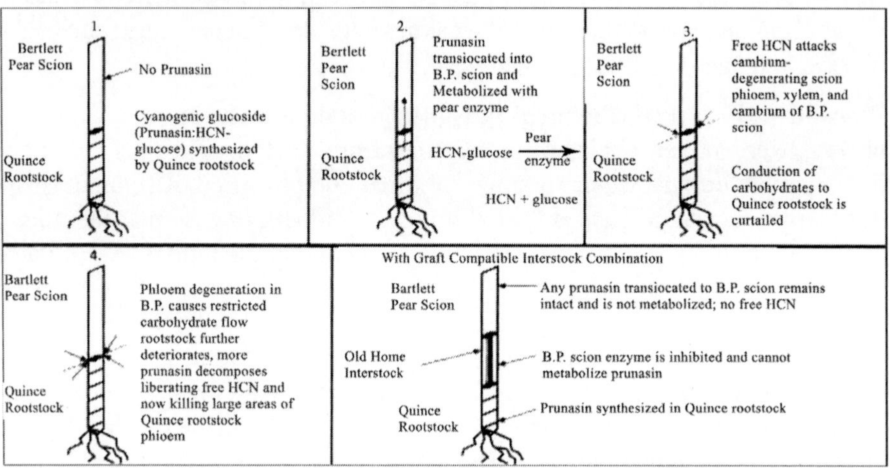

Fig. 1: Graft incompatibility in pear cv. Bartlett grafted on quince rootstock (*Source*: Hartmann & Kester, 2002, Plant Propagation: Principles and Practices)

Table 3: Important rootstocks and their influences on scion cultivars

S.No.	Fruit crop	Rootstock	Distinct influence on scion
1.	Mango	Totapuri Red Small	Dwarfing
		Vellaikollumban	Dwarfing
		Rumani,Olour	Dwarfing
		Kurukkan	Salt tolerance
		Creeping	Dwarfing
2.	Grape	Dogridge	Resistance to *Phylloxera*, nematodes and salts
		Salt Creek	Resistance to salt and nematodes
		St. George	Resistance to *Phylloxera* root louse
		Temple	Resistant against pierce's disease anthracnose and downy mildew
3.	Ber	*Zizyphus nummularia*	Dwarfing effect
4.	Guava	Pusa Srijan	Dwarfing effect on Allahabad Safeda scion cultivar
		Psidium pumilum	Used for the induction of dwarfing

5.	Citrus	Flying Dragon	Most dwarfing, highly suitable for high density planting
		Trifoliate orange	Deciduous, cold hardy, dwarfing, resistant to nematodes, resistant to most viral diseases
		Cleopetra mandarin	Most salt tolerant citrus rootstock
		Sweet orange	Resistant to tristeza,and exocortis
		Rangpur lime	Hardy rootstock, adaptable to various soil conditions and salt tolerant
		Rough lemon	Relatively tolerant to saline and calcareous soils
		Sour orange	Cold hardy, resistant to phytophthora root rot but highly susceptible to tristeza
		Citrus unshiu	Freeze tolerant
6.	Apple	M9	Dwarfing effect and highly suitable for high density planting
		M27	Ultra-dwarfing, suitable for high density planting
		Malus sikkimensis	Induces precocity in bearing
		M26	Better anchorage to scion cultivar
		M7	Semi-dwarfing, stronger and deeper root system
		MM111, MM106	Suitable for light sandy soils
		Northern Spy	Wooly aphid resistant
		EMLA series	Free from viruses
7.	Peach	Nemaguard	Resistant to nematode and crown gall
		GF-557	Nematode resistant
		GF-677	Drought tolerant, and high pH tolerant
8.	Pear	Quince C	Semi vigorous
9.	Plum	Pixie and St. Julien	Dwarfing rootstocks
		Marian 2624	Resistant to nematodes, crown gall, cold hardy and tolerant to high soil moisture
		Myrobalan B	Resistant to bacterial gummosis
10.	Almond	GF-557 and GF- 677	Tolerant to high soil pH
		Alnem1	Resistant to nematode

Role of hormones

It has been suggested by Sorce *et al.* (2002) and Koepke and Dhingra (2013) that plant hormones may play an important role in regulating the complex relationships between rootstock and scion. In vascular regeneration experiments, Aloni (1995) found that low concentrations (0.1 % w/w) of indole-3-acetic acid (IAA) stimulated phloem differentiation, whereas higher levels (1.0 %. w/w) induced xylem differentiation. An important class of substances in grafting experiments are the auxins released by the vascular strands of the stock and the scion, which induce the differentiation of vascular tissues, thus functioning as morphogenic substances (Aloni, 1987; Mattsson, 2003). Translocation from scion to stock observed by Shimomura and Fujihara, 1977, promoted successful graft unions in Cactus.

In grafted Cucurbita, Aloni *et al.* (2008) and Aloni (2010) showed that exogenous application of 1-naphthalene acetic acid (NAA), a synthetic auxin, caused root growth inhibition when applied to roots. Incompatible grafts affected root growth more than compatible ones. Hydroponically grown grafted plants were also treated with NAA directly to the roots, which caused root decay in incompatible grafts but less harm in compatible ones.

Three other observations corroborate auxin's involvement in incompatibility: (i) endogenous IAA analysis showed that the roots and stems of incompatible combinations contained higher levels of IAA than the same tissues of compatible ones; (ii) 2,3,5-triiodobenzoic acid, an auxin transport inhibitor, prevented root decay in incompatible combinations, but had only a slight effect on compatible combinations; and (iii) roots and shoots of incompatible grafts developed normally following blockage of basipetal IAA transport by partial stem gridling (Aloni *et al.,* 2008). After the graft connection is established, auxin produced in the stem is translocated downwards to the root, and once it reaches a threshold concentration, auxin triggers degradative processes causing root decay (Aloni, 2010).

Table 4: Rootstock/scion compatibility performance in different *Prunus* species and some interspecific hybrids

Rootstock type	**Pedigree**	**Compatibility performance**	**References**
Japanese plum/ peach	*Prunus salicina* x *Prunus persica*	Very good with plum and apricots, but not with peaches	Zaiger, 1982
	Prunus cerasifera and interspecific Hybrids	Performance differs substantiall depending on the genotype, 'Adara' rootstock confers compatibility even with cherry cultivars when it is used as an interstock	Zarrouk *et al.,* 2006
Peach	*Prunus persica* x *Prunus davidiana*	Wide range of compatibility with most peach and nectarine cultivars	Lang and Ophardt, 2000; Zarrouk *et al.,* 2006
Peach/ almond	*Prunus persica* x *Prunus dulcis*	Wide range of compatibility with most peach and nectarine cultivars	Lang and Ophardt, 2000; Zarrouk *et al.,* 2006
Slow-growing prunes	*Prunus domestica*	Wide range of compatibility with most peach and nectarine cultivars	Lang and Ophardt, 2000; Zarrouk *et al.,* 2006
Cherry	*Prunus avium* x *Prunus pseudocerasus*	Wide range of compatibility with most sweet cherry cultivars except 'Sam'and 'Van'	Romand Carlson, 1987

Cherry	*Prunus avium*	Incompatible with peach, nectarine and almond but compatible with all sweet cherry	Beckman and Long, 2003 and Lang, 2003
Cherry	*Prunus mahaleb*	Incompatible with several sweet cheery scion cultivars like 'Lapins', 'Chelan' and 'Tieton'	Long and Kaiser, 2010
Cherry	*Prunus cerasus*	Incompatible with several sweet cherry scion cultivars	Long and Kaiser, 2010
Cherry	*Prunus cerasus* x *Prunus canescens*	Wide range of compatibility with most sweet cherry cultivars	Long and Kaiser, 2010

References

Aloni, B. 1987. Differentiation of vascular tissues. *Ann. Rev. Plant Physiol.*, 38: 179-204.

Aloni, B., Karni, L., Deveturero, G., Levin, Z., Cohen, R. and Kazir, N. 2008. Physiological and biochemical changes at the rootstock-scion interface in graft combinations between cucurbita rootstocks and melon scion. *J. Hort. Sci. Biotech.,* 83: 777-783.

Aloni, R. 1995. The Induction of Vascular Tissues by Auxin and Cytokinin. In P.J. Davies (ed.) Plant Hormones. Kluwer Academic Publishers. Dordrecht. The Netherlands. pp.531-546.

Aloni, R. 2010. The Induction of Vascular Tissues by Auxin Plant Hormones. In P.J. Davies (ed.) Springer, Dordrecht, The Netherlands. pp. 485-518.

Beckman, T.G. and Lang, G.A. 2003. Rootstock breeding for stone fruits. *Acta Hort.*, 622: 531-551.

Bose, T. K., Mitra, S. K., Sadhu, M. K. and Das, P. 1986. Propagation of Tropical and Subtropical Horticultural Crops. 2nd Edition, Naya Prakash, Calcutta.

Chadha, K. L. 2019. Handbook of Horticulture. 2nd Edition (Revised). Directorate of Knowledge Management in Agriculture, Indian Council of Agriculture Research, Pusa, New Delhi.

Hartmann, H.T., Kester, D.E., Davies, F.T. and Geneve, R.L., 2002. Plant Propagation. Principles and Practices. 7th ed. Prentice Hall, Upper Saddle River, NJ, p.849.

http://ecoursesonline.iasri.res.in/mod/page/view.php?id=96813

https://www.biotecharticles.com/Agriculture-Article/Graft-Incompatibility-in-Fruit-Crops-PDF-3887.html

https://www.slideshare.net/pawannagar8/principle-amp-different-method-of-cutting-amp-layering

Koepke, T. and Dhingra, A. 2013. Rootstock scion somatogenetic interactions in perennial composite plants. *Plant Cell Report*, 32: 1321-1337.

Lang, G.A. and D. Ophardt. 2000. Intensive crop regulation strategies in sweet cherries. *Acta Hort.*, 514: 227-234.

Long, L.E. and C. Kaiser. 2010. Sweet Cherry Rootstocks. Pacific Northwest Extension, 619: 8.

Mattsson, J. 2003. Auxin signaling in Arabidopsis leaf vascular development. *Plant Physiol.,* 131:1327–1339.

Nath, V., Saroj, P. L., Singh, R. S., Bhargava, R. and Pareek, O. P. 2000. *In-situ* establishment of ber orchard under hot arid ecosystem of Rajasthan. *Indian J. Hort.,* 57(1): 21-26.

Rom, R.C. and Carlson, R.F. 1987. Rootstocks for Fruit Crops. Wiley, New York, USA.

Sharma, R. R. 2002. Propagation of Horticultural Crops: Principles and Practices. Kalyani Publishers, New Delhi.

Shimomura, T., and Fujihara, K. 1977. Physiological study of graft union formation in Cactus. II. Role of auxin on vascular connection between stock and scion. *J. Japan Soc. Hort. Sci.,* 45(4): 397-406.

Singh, J. 2014. Basic Horticulture. Kalyani Publishers, New Delhi.

Sorce, C., Massai, R., Picciarelli, P., and Lorenzi, R. 2002. Hormonal relationships in xylem sap of grafted and ungrafted *Prunus* rootstocks. *Sci. Hort.*, 93: 333-342.

Zaiger, C.F. 1982. Interspecific Rootstock Tree 4- G-816. In Patent A01H 5/03, USA.p.3.

Zarrouk, O., Gogorcena, Y., Moreno, M.A. and Pinochet. J. 2006. Graft compatibility between peach cultivars and *Prunus* rootstocks. *Hort Sci.,* 41(6):1389-1394.

6

Role of Plant Growth Regulators in Propagation of Fruit Crops

M.L. Jat, R.K. Jat and O.P. Kumawat

Introduction

The term plant growth regulator (PGR) generally refers to organic compounds other than nutrients that influence plant growth and development at low concentrations. According to practical definitions, plant growth regulators are either natural or synthetic compounds applied directly to plants to alter their life processes or their structure in order to improve quality, increase yields, or to facilitate harvesting. The term plant hormone, when correctly used is restricted to naturally occurring plant substances. This fall into five classes: auxins, gibberellins, cytokinins, inhibitors and ethylene (gas). Plant growth regulators include synthetic and naturally occurring hormones.

In general, plant growth regulators are commonly used in techniques of plant propagation, such as seed germination, dormancy, cutting, grafting, tissue culture, and micropropagation.

Main functions

Auxins: Increase the size of plant cells.

Giberellins: May stimulate cell growth, cell division, and cell elongation.

Cytokinins: Promote plant cell division.

Inhibitors: Plant hormones that inhibit or delay physiological or biochemical processes.

A. Rooting in hard to root stems cuttings

Historically, plant growth regulators were used to promote or accelerate the rooting of cuttings. Many types of PGRs are used in the form of liquid solution and paste to initiate rooting. Concentration of plant growth regulators is determined by the plant species, types of cuttings and method of application. For this purpose, indole butyric acid (IBA) is the most popular and best chemical to use, which is decomposed relatively slowly by the auxin-destroying enzyme system in plants. Due to its slow movement in the plant, this compound is also retained close to the site of application. Another highly active auxin commonly used for root promotion is naphthaleneacetic acid (NAA). NAA is more toxic than IBA, so treated plants face a greater risk of injury. NAA and IBA both work effectively as a rooting stimulant. When used at low concentrations, phenoxy compounds, such as 2,4-D, 2,4,5-T (trichlorophenoxyacetic acid), can encourage root formation. The type of root systems formed depends on the growth regulator used. They tend to produce bushy, stunted and thickened roots systems. Commercially, IBA is used as a rooting hormone in many types of horticultural and forest plants, such as apple, peach, plum, ficus, kiwifruit, pomegranate, rose, tea plants, winged bean plants, rhododendrons, egg plants, *etc.*

Methods of application: There are some methods of applying growth regulators to stem cuttings for the induction of roots: 1. Quick dip method, 2. Prolonged soaking method (dilute solution soaking method), 3. Powder method and 4. Lanolin paste method.

1. Quick dip method (concentrated solution dip)

As part of the quick dip method, a concentrated solution of auxin varying from 500 to 10,000 ppm (0.05 to 1.0 percent) is prepared in aqueous solution or 50 % alcohol, and the cut end (0.5 to 1 cm) is submerged for a short period of time (usually 3 to 5 seconds, sometimes longer). Afterwards, the cuttings are inserted into the rooting medium. Cuttings are best dipped in a bundle rather than individually. Many propagators prefer a quick dip application over a talc application because it gives consistent results and is easier to apply. At the end of the day, change the solution rather than pouring it into the stock solution. In open spaces where evaporation is high on hot days, it is better to discard the old solution and add fresh water several times during the day. Stock solution that contains a high percentage of alcohol will retain their activity almost indefinitely if kept clean. When handling these compounds, it is recommended to wear rubber or plastic gloves.

2. Prolonged soaking method

As part of this procedure, the basal ends of cuttings are soaked in diluted solution (10 to 500 ppm) for up to 24 hours just prior to being planted in the rooting medium. The concentration varies from 20 ppm in the case of easily rooted cuttings to 200 ppm in the case of difficult to root species. It is recommended that the cuttings be held at 20° C for the entire soaking period. However, they should not be placed in the sun. The technique is generally slow, cumbersome, and not widely used. There may be variability in the results due to the long soak time, and environmental changes may occur during the soak.

3. Powder method

In this procedure, the cuttings' basal ends are treated with the growth regulators in a carrier usually a clay or a talc. The concentration of active ingredients in the inert carrier is between 500 to 1000 ppm. Talc preparations are convenient and easy to use. It may be difficult to achieve uniform rooting, due to the length of time it takes the talc to adhere to the base of the cutting, the amount of moisture at the base of the cutting, the texture of the stem (*i.e.*, coarse or smooth) and loss of the talc during insertion of the cutting into the propagation medium. At comparable concentrations, talc formulations tend to be less effective in solution than IBA. These formulations are still popular with nurserymen, such as Seradax-A, B, and Rootone.

4. Lanolin paste method

During the preparation of hormonal pastes, the required dosage of hormone is weighed precisely and dissolved in a few drops of alcohol. The required amount of lanolin is weighed and heated slightly in a beaker under gentle flame. After the lanolin is slightly liquidized, the dissolved hormone is added. All ingredients are dissolved, mixed thoroughly, and allowed to cool. After cooling, the paste is ready for use. In the case of a layer or stool, growth regulators are applied to the girdled portion of a layer or stool in lanolin paste for inducing rooting.

B. Seed germination

Gibbrellic acid is positively correlated with seed germination in many fruit plant species. Presoaking the seeds help to break the dormancy and enhances the seed germination. *E.g.*, Guava, ber, potato and sugarcane.

C. Bud sprouting

Application of gibbrellic acid, IBA, cytokinin, ehtrel and thiourea stimulate bud sprouting. The sprouting buds can be used in grafting, budding and other vegetative propagation techniques.

D. Breaking the dormancy

Plant hormone affects the seed germination and dormancy by acting on different parts of seeds. Embryo dormancy is characterized by ABA levels that are high compared with GA levels, while seed levels are low compared with ABA. GA, ethrel and NAA are used in breaking dormancy in seeds and buds.

The treated seeds can be sown to propagate plants sexually. The sprouted buds may be used for budding, grafting, or other types of vegetative propagation.

E. Micropropagtion

The use of growth regulators in the culture media is critical to the effectiveness of the micropropagation technique. Growth regulators regulate the growth and developmental processes. These are the most important components in starting the regeneration process in tissue culture. Explants do not react well to culture conditions without growth regulators in the majority of *in-vitro* investigations. *In-vitro* growth and differentiation of plant tissues is regulated by an interaction between auxins and cytokinins. The following reviews are concerned with the role of growth regulators in woody plant micropropagation.

1. *In-vitro* seed germination and shoot elongation

In-vitro propagation is used to propagate a number of woody plants. GA_3 is a growth regulator that promotes seed germination and shoot lengthening. Several workers implemented GA_3 in their culture media and found it to be beneficial (Isogai *et al.*, 2008; Balaraju *et al.*, 2011; Joseph *et al.*, 2011 and Al-Safadi and Elias, 2011). GA_3 was shown to be the most efficient growth regulator for shoot elongation in *Melastoma malabatricum*, according to Ghimire *et al.* (2016). GA_3 was the most effective for shoot elongation. But in *Pyrus boissieriana,* GA_3 negatively affected number and length of shoots (Zakavi *et al.*, 2016).

2. *In-vitro* shoot regeneration

It is generally true that basal nutrient medium without growth regulators is not as effective in stimulating shoot buds in woody plants. Similarly no shoot buds developed in *Crataeva nurvala* (Walia *et al.*, 2003) and *Cinnamomum camphora* (Sharma and Vashistha, 2010) on basal medium. A growth regulator applied exogenously has variable effects that are influenced by its type, its concentration and nature of explants. There are two methods for *in- vitro* shoot regeneration in woody species: direct and indirect organogenesis.

References

Al-Safadi, B. and Elias, R. 2011. Improvement of caper (*Capparis spinosa* L.) propagation using *in-vitro* culture and gamma irradiation. *Sci. Hort.*, 127: 290-297.

Balaraju, K., Agastian, P., Ignacimuthu, S. and Park, K. 2011. A rapid *in-vitro* propagation of red sanders (*Pterocarpus santalinus* L.) using shoot tip explants. *Acta Physiol. Plant*, 33:2501-2510.

Ghimire, B.K., Seong, E.S., Nguyen, T.H., Yu, C.Y., Kim, S.H. and Chung, I.M. 2016. *In- vitro* regeneration of *Melastoma malabatricum* Linn. through organogenesis and assessment of clonal and biochemical fidelity using RAPD and HPLC. *Plant Cell Tiss. Organ Cult.*, 124: 417-529.

http://ecoursesonline.iasri.res.in/mod/page/view.php?id=96847

Isogai, S., Touno, K. and Shimomura, K. 2008. Gibberellic acid improved shoot multiplication in *Cephaelisi pecacuanha. In Vitro Cell. Dev. Biol. Plant*, 44:216-220.

Joseph, N., Siril, E.A. and Nair, G.M. 2011. An efficient *in-vitro* propagation methodology for annatto (*Bixa orellana* L.). *Physiol. Mol. Biol. Plants,* 17:263-270.

Sharma, H. and Vashistha, B. D. 2015. *In-vitro* plant regeneration through callus in giloy (*Tinospora cordifolia* (Willd.) Miers ex Hook f & Thoms.). *Indian J. Sci.,* 12(34): 59-68.

Usha, K. and Nimishab, S. 2014. Use of plant regulators in horticultural crops. *Biotech article.*

Walia, N., Sinha, S. and Babbar, S.B. 2003. Micropropagation of *Crataeva nurvala. Biol. Plant*, 46:181-185.

Zakavi, M., Askari, H. and Irvani, N. 2016. Optimizing micropropagation of drought resistant *Pyrus boissieriana* Buhse. *Physiol. Mol. Biol. Plants*, 22(4): 583-593.

7

Plant Tissue Culture and Micropropagation

Kapil Mohan Sharma

Introduction

Among the various propagation techniques, plant propagation through tissue culture has gained immense importance over the years and for a few crops it has become the major method of propagation. Plant tissue culture can be defined as the ability to grow and maintain the plant organs *viz.*, embryo, shoot, root, anther, flower and their respective tissue in a culture media. While micropropagation can be defined as the development of a new plant using plant tissue culture techniques. These plants are grown under *in-vitro* conditions, *i.e.*, culturing the plant in a vessel (*e.g.*, test tube) under a controlled condition. Since, the whole process is a lab-based work, there are chances of contamination, thus, the growing condition needs to be aseptic (growing without microbial contamination).

The idea of the development of the micro-propagated plant is based on the totipotency of the plant, *i.e.,* the plants' potential to regrow itself, even from its tissues. The term 'totipotency' was coined by a German plant physiologist Haberlandt (1902) and is considered as the father of tissue culture technique. This was utilized for the development of micro-propagated plant by Prof. G. Morel in 1958 during his experimentation in developing a virus free planting material of dahlia and orchids. This has opened a wide range of scope to develop methodology to propagate true to type plants of commercial importance (banana, pineapple, *etc.*), rare plants (endangered medicinal plants, orchids, *etc.*) or for the plants in which there is no or limited opportunity to propagate plant vegetatively (coconut, date palm, *etc.*). Today, protocol for plant tissue culture is available for more than a thousand species and many of them are now also commercialized. Currently, a major use of this technique is done in the countries like India and China.

Plant tissue culture-based industries in India

Currently, India is one of the major countries contributing in the micropropagation of its crops. The major rise in the production of micro-propagated plants is observed in the recent years majorly due to the availability of the cheap labour. The rise of the micropropagation industry in India started with 1987-1988 with the setup of India's first commercial plant unit in Cochin. The number of registered industries has risen since then and currently 92 such commercial tissue culture units are working (Table 1) distributed in different parts of the country. The major crops which are successful and are in high demand in tissue culture raised plants are banana, pineapple, strawberry, lily, roses, orchids, ginger, turmeric, cardamom, black pepper, gerbera, pomegranate, female papaya, vanilla, coffee, tea, eucalyptus, teak, rosewood, *etc.*

Table 1: Number of Commercial Tissue Culture Production Units (TCPUs in India (as on December, 2021)

State	Number of TCPUs	State	Number of TCPUs
Andhra Pradesh	3	Madhya Pradesh	5
Bihar	1	Orissa	2
Chattisgarh	5	Punjab	2
Gujarat	18	Rajasthan	1
Haryana	3	Tamil Nadu	9
Himachal Pradesh	1	Telengana	5
Karnataka	10	Uttar Pradesh	3
Maharashtra	22	West Bengal	2

Source: National Certification System for Tissue Culture Raised Plants (NCS-TCP)

Advantages of micropropagation

1. **Large number of planting materials:** The major advantage of the micropropagation is its ability to produce a large number of planting material in a short period of time, which is not possible with the traditional methods.

2. **Disease free material:** Since, the plantlets are developed through the meristematic part, it is devoid of any contamination especially the viral load which is very difficult to eradicate in the other traditional methods.

3. **True to type material:** The planting materials produced through micropropagation are uniform and true to type in nature and possess the same characters as that of its mother plant.

4. **Small size and portable:** Since, the planting material is small in size, and initially in test-tubes, they can be easily transferred from one place to another, and comparatively needs smaller space, making it cheaper during transportation.
5. **Production throughout year:** Since, the planting material are produced in a controlled condition, they are produced throughout year irrespective of the season.
6. **Useful in dioecious plant:** Crops like date palm and papaya, propagating only female plants can be easily done though micropropagation method. It reduces the excess labour needed to eradicate unwanted male plants.
7. **Useful in hard to root plants:** It is suitable for hard to root plants where most vegetative methods are not successful and often results in failed development of propagules.
8. **Overcome seed dormancy:** Crops which have seed dormancy and are hard to germinate can be propagated through this technique.
9. **Easy to popularize new varieties:** To popularize a new variety needs planting material, and such demand and supply gap can easily be eliminated through micropropagation.
10. **Smaller breeding cycle:** To breed and take trails of a new cultivar, micropropagation allows reduce breeding cycle.
11. **Helpful in getting intergeneric hybrids:** Often difficult to cross, of intergeneric crosses, whose success rate is very low can be extracted and multiplied though this technique and create rare hybrids.
12. **Germplasm protection:** Through this technique, some rare germplasm can be conserved and propagated. They can also be cryopreserved and repropagated through micropropagation.

Disadvantages of micropropagation

1. **High cost:** The production of tissue culture plantlets needs sophisticated labs and infrastructure to raise quality planting material, which needs high initial investment which might not be feasible for many nurserymen.
2. **Contamination:** Although the planting material are grown in controlled condition to reduce contamination, but if any contamination got introduced in such facility it may contaminate other propagules in very short span and very difficult to re-eradicate the contaminant.

3. **Somatic variation:** Tissue cultured raised plants are true to type but due to irresponsible usage of growth regulators, poor planting material or management practices there may be chances of somatic variation, where the explant have different character from its mother plant.
4. **Poor establishment:** The plant raised in controlled condition and poorly hardened and when exposed to harsh field condition, they may show mortality.
5. **Need of skilled manpower:** The lab based work needs skilled manpower and technical guidance, devoid of which, it is difficult to raise plants.

Principles of micropropagation

The development of *in-vitro* plants is based on the "totipotency" hypothesis by Gottlieb Haberland in 1902. The hypothesis proposes that each cell of the plants has genetic information in them and are has potential ability to produce a perfect plant. The differentiated cells in the plants can dedifferentiate themselves and re-enter the cell cycle, proliferate and regenerate tissues and organs. This ability of plant provided opportunities to produce a large number of plants. This can be done majorly by two methods, somatic embryogenesis or organogenesis. Organogenesis is the process by which a whole plant or a plant part (organ) can be formed, while, in somatic embryogenesis, initially the cells similar to zygotes are formed and then the entire plant is regenerated. This regeneration allows plant to propagate clones through cutting and grafting. To promote regeneration, over the years, several phytohormones, explant type, donor plant's physiological properties, their nutritional uptake and distribution pattern, potential of a plant part to dedifferentiate and redifferentiate, *etc.* were studied.

Organogenesis

Organogenesis is the initiation and development of plant part or organ. In plant tissue culture it is important to regenerate the plant from the culture. In *in-vitro* culture it can be divided into two distinct phases: dedifferentiation and redifferentiation. Dedifferentiation starts shortly after the explant tissues are isolated, cells started dividing and forms mass of undifferentiated cells called callus. Redifferentiation starts after the formation of callus with the development of tissues called organ primordia which were differentiated from a single or a group of callus cells. During this stage the plants were aligned for polarity along the longitudinal axis with the development of meristematic tissues. These cells further specialize as they differentiate. Vascular system is formed connecting the new organs to their parent explant.

Somatic embryogenesis and synthetic seed production

Somatic embryogenesis is the development of the embryo *i.e.*, distinct shoots and roots from vegetative cells instead of zygote (fusion of male and female gametes). This method is now been used commercially as well as by breeders to develop synthetic seeds, also called as synseeds. The synthetic seeds are developed by the encapsulating the somatic embryo into a synthetic seed coat which is suitable for sowing. These are useful in the development of variants using somaclonal variation, genetic transformation, development of plantlets using haploids, development of cybrids and in germplasm conservation through cryopreservation. The major advantage of synseeds is:

- Easy clonal and seed propagation technique.
- Easy to induce agricultural chemicals and microorganisms.
- Uniform yield.
- Helps to conserve high value germplasm.
- Reduces time period.

Techniques in micropropagation

The development of micropropagation can be made by the following methods or techniques

Meristem culture: In this method, meristems are used as explants for culturing the plant. The meristematic tissues are the type of tissues which can divide and differentiate. Such tissues are present in the tip of the shoots and roots. This method is best for virus elimination and plant breeding purposes.

Callus culture: In this method, callus cells are used for the culturing the plant. The callus is an undifferentiated mass of tissues. The callus development is initiated by placing tissues in the growth culture media under favourable conditions. However, callus shows high biochemical variability and growth rate is comparatively slow.

Suspension culture: Suspension culture is developed by multiplication of single cells or aggregates in suspension agitated with aerated and sterile liquid medium. This includes both batch culture and continuous culture methods. The most common problem is formation of clumps and failure of cells to separate after division.

Embryo culture: In these methods, embryo is isolated from immature or mature embryos and culture grown in a suitable media and conditions. This process is very helpful in the plants were endosperm degenerate followed by culturing the immature hybrid embryo. This method is called embryo rescue and has massive role in crop improvement.

Protoplast culture: Protoplast is a living cell without cell wall (naked cell). Removing of the cell wall using chemical, mechanical or enzymatic process and culturing them in suitable media is called protoplast culture. Primarily, leaves are the common explant used followed by root tips and embryo.

Micro grafting: It is a *in-vitro* grafting technique in which meristem or shoot tip explant are placed on the decapitated rootstock and are grown under suitable media and conditions as micro-propagated cultures.

Stages of micropropagation

Stage I: Establishment: This stage deals with the handling of the stock plant/mother plant from which the planting material is to be made. The explant used should be obvious devoid of any pathogen and for which "virus free" or pathogen indexed plant should be used. Choice of the type of explant depends on the method or purpose of the propagation. The explants are disinfected and pre-treated for removal of any additional toxicity. The explants are transferred to liquid based media where they are allowed to induce callus.

Stage II: Shoot multiplication: The next stage followed is subculturing where the callus formed in the last stage are transplanted into a new media to develop multiple shoots. It is the most important step to decide the commercial utility of the shoot. The process is repeated multiple times to develop large number of micro shoots.

Stage III: Root multiplication: In this step, developed shoots are made to regenerate *in-vitro* rooting after transferring them into an agar based media. This helps to make the complete plantlets.

Stage IV: Acclimatization: The last step of the process is to transfer the completely made plantlets into a hardening units like greenhouse, shade nets to acclimatize into the outer environment. This stage is also called the hardening stage and is important before the final transfer of the plantlets in the open condition.

Factors affecting micropropagation

The development of the plantlets is a skilled work; however, a large number of factors influence the success of the development of the propagules. The factors influencing the *in-vitro* propagation are enlisted below:

i. **Media of propagation:** Different media are used in different situations *e.g.*, White medium, Nitsch and Nitsch medium, B5 medium, MS medium, *etc.*

ii. **Type of explant:** Based on the plant type plant parts like leaf, petiole, cotyledonary leaf, hypocotyle, epicotyle, embryo, internode, root explant, *etc.* are used as explant.

iii. **Genotype:** It varies because of their internal level of hormones.

iv. **Source of explant:** *In-vitro* explant or *in-vivo* explant.

v. **Orientation of explant:** Horizontal has higher efficiency than vertical.

vi. **Mineral nutrition:** Optimization of levels of nutrition, mostly N, P and K is needed; both higher and lower level may affect growth.

vii. **Carbon source:** Carbon source is needed to compensate photosynthesis. Sucrose is most commonly used carbon source.

viii. **Growth regulators:** Ratio of various growth hormone *i.e.*, auxin and cytokinin is vital for success of plant growth. Higher level of auxin to cytokinin induces roots, while higher level of cytokinin induces shoots.

ix. **Gelling agents:** Both solid and liquid media are used. Liquid media are preferred for homozygous nature, cheap and easy to produce. Agar media is traditionally used as semi-solid media for plant tissue culture.

x. **Gas exchange and relative humidity:** Levels of carbon dioxide, oxygen and ethylene are the important gases studied to understand growth. Relative humidity of 98-100 % is needed for proper growth.

xi. **Light:** Wavelength, flux density and duration of light exposure influences the growth.

xii. **Temperature:** Moderate temperature between 20°C and 27°C is most optimal temperature for growth.

Genetic fidelity testing and virus indexing

To maintain the standard of the plant, the plants are screened for their genetic fidelity and virus indexing and specified labs are now also available where such facilities are available. To grow the plant through micropropagation, several combinations of plant growth regulators are used in various combination. Such employed methods may also alter the genetic matter of the plantlets called as somaclonal variation. This might be in the form of the DNA methylation, chromosome rearrangements, and point mutations. Thus, it is important to check the genetic fidelity of the plant to prove that they are true to type to the main plant and maintains clonal stability to maintain characteristics of the mother plant. This is done mostly by using Inter Simple Sequence Repeat (ISSR) and Random Amplified Polymorphic DNA (RAPD).

Virus indexing is the procedure to determine the presence or absence of the viruses in the given planting material. Primarily, this is done by grafting the symptomless plant (plant to be tested) onto an indicator plant (very susceptible to the virus) and their respective observations were made. For example, tristeza virus is a common problem in *Citrus*, however, *Citrus aurantifolia* are highly susceptible to tristeza and are used as an indicator plant (rootstock) for virus indexing of the tristeza virus. However, in recent days ELISA (Enzyme Based Immuno Sorbent Assay) based test are preferred and are widely used for early detection of virus in the propagules.

National Certification System for Tissue Culture Raised Plants

Before commercializing the product, both genetic fidelity test and virus indexing are done to confirm the quality parameters of the products. Currently, five major accredited test laboratories are working under National Certification System for Tissue Culture Raised Plants (NCS-TCP) (Table 2). National Certification System for Tissue Culture Raised Plants (NCS-TCP) is a unique quality management system established by Department of Biotechnology (DBT), Government of India, under the "Seeds Act, 1966" of the Ministry of Agriculture and Farmers' Welfare for ensuring the production and distribution of quality tissue culture planting materials.

The tissue culture plant production facilities based on compliance with technical capabilities, infrastructure, package of practices and documentation/ record keeping get recognition under this quality management system. Once the tissue culture production facility is recognized they become eligible to get their tissue culture raised planting material certified from Accredited Test Laboratories (ATLs). The certified batches of tissue culture raised plants are provided with certification labels enabled with barcode. This barcode provide end users to trace back history of plants from where the tissue culture raised planting material are derived.

Table 2: Accredited Test laboratories (ATLs) under NCS-TCP

S. No.	Accredited Test Laboratories
1.	University of Agricultural Sciences, GKVK, Bangalore
2.	ICAR-Central Potato Research Institute, Shimla
3.	ICAR-National Research Centre for Banana, Tiruchirapally
4.	ICAR- Indian Institute of Sugarcane Research, Lucknow
5.	Vasantdada Sugar Institute, Pune

The minimum quality standards for the development of the planting material are as follows:

i. Effective sanitation practices including insect and disease monitoring and prevention must be adhered to.

ii. No field produced plants can be grown in the protected environment (greenhouse/ polyhouse) along with tissue cultured plants.

iii. Varieties must be separated by physical barriers (such as proper tagging), which will prevent varietal mixture.

iv. Before dispatch to the farmers, the tissue cultured plants growing in the nursery should be tested for the absence of the viruses such as apple mosaic virus, apple chlorotic leaf spot virus, and clonal uniformity. For establishing clonal fidelity, the sample size should be 0.1 % of the batch size with a minimum of 10 plants.

v. If testing performed by an accredited laboratory reveals the presence of banned viruses, fungus or bacteria the tissue cultured plants should be dispatched from the premises of the production lab and the entire material should be destroyed.

vi. The concerned laboratory/agency producing the tissue culture raised material should issue a certificate to the effect that the ATC have been produced as per guidelines.

vii. The agency producing ATC will follow the standard labelling procedure (Figure 1).

NCS-TCP

Certified Tissue Culture Raised Quality Plants/Propagules

Certificate of Quality No.: ....

Label No.: ....

Name of Production Facility: ....

Botanical Name: ....

(Common Name): ....

Certification No. and validity of Certificate of Recognition: ....

Variety: ....

Batch No. & Batch Size: ....

Contact person and Designation: ....

Stage of Tissue Culture Plants:

Address with phone number: ....

In agar [ ] Ex-agar [ ] Hardened [ ]

Bar Coding :

Date of Issue: ....

Name/Sign/Stamp of ATL with date: ....

Fig. 1: Plant Label as per the National Certification System for Tissue Culture Raised Plants (NCS-TCP)

Packaging and transport of micro-propagules

A standard package of practices has been made for tissue culture propagation unit. It includes a few major steps as presented in Figure 2.

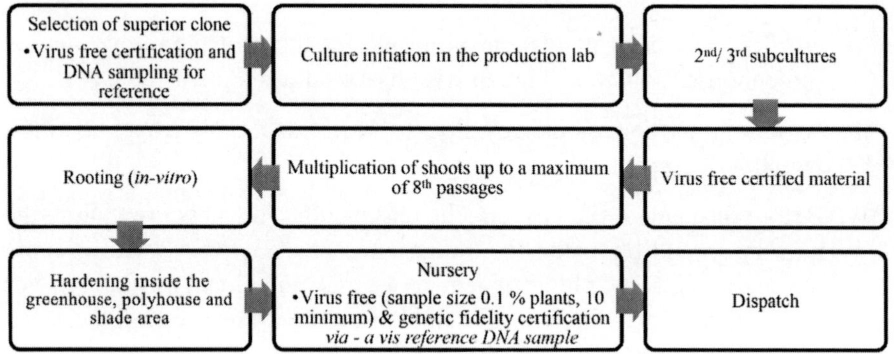

Fig. 2: Steps in the development of micro-propagules

The plants are packed either after hardening or before hardening. Before hardened plants are convenient to transport and avoid loss of plants. They also prevent contamination due to soil borne diseases during hardening. The method of packaging depends on the location and objective of the shipment. The following packaging materials are widely used in the plant tissue culture.

- Test tubes plugged with cotton
- Foldable bags
- Plug trays
- Semi-permeable plastic bags
- Him-burg boxes
- Thermocoal
- Thermocoal with refrigerated condition
- Corrugated fiber boxes
- Pots

The products packed are transported through various cargo methods. The pre-hardened plants are transported under refrigerated condition. The plantlets in liquid media are transferred to solid media to avoid damage. It is essential to maintain optimum temperature and relative humidity during transport for better survival rate. Further, proper management of phytosanitary standards are also required for quality management. During packaging and transport, proper labelling and certification is a prerequisite.

References

Anonymous, 2022. National Certification System for Tissue Culture Raised Plants (NCS-TCP).

Hartmann, H. T., Kester, D. E., Davies, F. T. and Geneve, R. L. 2011. Hartmann and Kester's Plant Propagation: Principles and Practices (8th edition). Pearson Prentice Hall, New York. p. 914.

Sharma, R. R. and Krishna, H. 2013. Textbook of Plant Propagation and Nursery Management. IBDC Publishers, Lucknow. p.194.

URL: https://dbtncstcp.nic.in/. Accessed on 01/04/2022.

8

Plant Propagation Structures

Jitendra Singh Shivran, Rajkumar Jat, Pooja Sharma and Mukesh Chand Bhateshwar

Introduction

After the green revolution, India achieved significant strides in agricultural output, but productivity is still poor owing to climate change and the interference of old agriculture techniques. In order to fulfill the world's food demand in 2050, global output will have to grow by 70 % (FAO, 2020). Man has devised technical techniques for growing crops round the year.

Crop propagation is a natural process that is enhanced *via* the use of modified and regulated settings. These conditions are altered to guarantee that the crop in question grows to its full potential. The purpose of propagation structures is to build environments that can be changed and controlled.

Different propagation structures

Propagation is carried out in a variety of structures that is deliberated below:

1. Greenhouse

Horticulturists have long used greenhouses to stimulate plant growth (Beytes, 2003). Glass or plastic film, as well as transparent and translucent materials, are used to cover greenhouse buildings. Now, there are two types of practice environments: a completely controlled environment and a partially controlled one. This method of growing plants in greenhouses is quite important, especially in locations where climatic conditions are always on the severe side and where rainfall is abundant. For agricultural cultivation, several types of greenhouse buildings are available. Because all types of greenhouse structures have benefits and disadvantages for certain applications, no one form of greenhouse is regarded to be the best. As a result, many greenhouse designs based on usefulness, shape, material, and construction are available to fulfill the demands of different people (Dalai *et al.*, 2020).

Types of greenhouses

A. Types of greenhouse based on cost investment

1. Low technology greenhouses

Only a tiny percentage of farmers utilize low-tech buildings to produce their crops. The overall height of the low-tech greenhouses is less than three meters. Tunnel homes, the most prevalent kind of low-tech greenhouse, have poor ventilation due to their lack of vertical walls. Because little or no automation is utilized, this style of construction is quite cheap. However, as compared to open field farming, this sort of structure offers significant crop yield benefits. The growing environment still limits agricultural potential, and crop management is challenging. Low-tech greenhouses usually provide a suboptimal growth environment, resulting in lower yields and no reduction in the frequency of pests and diseases. Low-tech greenhouses allow for a more affordable entry into the farming sector.

2. Medium technology greenhouses

In medium-technology greenhouses, vertical walls are common. An average greenhouse has a height of more than 2 meters, but less than 4 meters, with a total height of less than 5.5 meters. For greater ventilation, these greenhouses include either a roof or a side wall. They also offer medium automation and are often coated in single or double-skin plastic film or glass. Medium-level greenhouses offer an acceptable economic and environmental base for the farming business, since they strike a balance between cost and production. Crops grown in medium-level greenhouses are more efficient than those grown on open fields. Hydroponic systems make water consumption more efficient. Although there is more possibility to utilize non-chemical pest and disease control techniques, greenhouse horticulture's full potential is challenging to achieve.

3. High level greenhouses

The height of the high-level greenhouses wall must be at least 4 meters, and the roof peak must be at least 8 meters above ground level. These buildings provide high agricultural yield while still being environmentally friendly. This building has enough roof ventilation as well as side wall vents. As cladding materials in high-level greenhouses, plastic films (single or double), polycarbonate sheets, and glass are used. There are environmental controls that are automated.

These buildings provide significant economic and environmental sustainability potential. Pesticide use has also decreased. High-tech buildings are typically stunning to look at, and they are increasingly being used in agriculture possibilities across the world. While these greenhouses are expensive, they provide an environmentally sustainable and highly productive option for the modern fresh produce industry. Wherever possible, investment decisions should prioritize the installation of high-tech greenhouses.

B. Greenhouse type based on shape

1. Lean-to type greenhouse

Lean-to greenhouses are those that have one or more of their sides against the side of an existing greenhouse building. The current greenhouse structure's roof has been expanded with covering material, and the exposed space has been adequately covered. The entire building should face south, which is the optimum orientation for getting enough sunlight. Typically, this form of greenhouse consists of single- or double-row plant benches that are 7 to 12 feet wide and as long as the structure they are attached to. The advantages of a lean-to greenhouse are that it is generally the cheapest construction and is located near a source of energy, water, and heat. This style of greenhouse maximizes the utilization of natural light while reducing the need for roof supports. Among the drawbacks are limited space, limited light, limited ventilation, and limited temperature control.

2. Even span type greenhouse

Even span greenhouses are built entirely on ground level and have two roof slopes of equal width and pitch. When the greenhouse is tiny and linked to a home at one gable end, this form of greenhouse building is utilized. The number of plant benches in this sort of greenhouse is limited to two or three rows.

Despite the fact that an even-span greenhouse construction is more expensive than a lean-to structure, even-span greenhouses offer greater design flexibility and can accommodate more plants. Because of the size of the greenhouse structure and the large amount of exposed glass, it will be more expensive to heat. In comparison to a lean-to construction, the design has a superior shape for air flow inside the structure to maintain consistent temperatures during the winter season. Two side benches, two walkways, and a large central bench are all included. In different parts of India, several types of single and multiple span constructions are available. Single span structures range in span from 5 to 9 meters, with a length of 24 meters and a height of 2.5 to 4.3 meters.

3. Uneven span type greenhouse

This style of greenhouse has an uneven spread and is ideal for hilly or undulating terrain. The structure's roofs have different widths, which corresponds to the name of the construction and allows it to adjust to hillside slopes. This sort of greenhouse is rarely utilized since it is not automatable.

4. Ridge and furrow type greenhouse

In this type of greenhouse, two or more A-frame structures are connected along the length of the eave. The eave serves as a gutter or furrow that collects rainwater and snow melt. A single large interior has been created as a result of the disappearing side wall between the greenhouses. Using greenhouse structures reduces labour, lowers automation costs, improves personal management, and lowers fuel consumption since there is less wall area through which heat escapes. Because snow does not roll off the roofs like it does on individual free-standing greenhouses, it must be factored into the frame requirements of these greenhouses. Canada and northern European nations successfully utilize ridge and furrow greenhouses despite heavy snowfall.

5. Saw tooth type greenhouse

The saw-tooth structure of the greenhouse allows for natural ventilation, similar to ridge and furrow greenhouses (Fig. 1). The saw tooth vent may be opened to enable continuous airflow to lower the internal temperature or closed to optimize the climate management of the growth space. In addition to the side ventilation, the roof ventilation alone provides 25 % of the overall ventilation of the covered area. The arches' form enables for a lot of light to pass through.

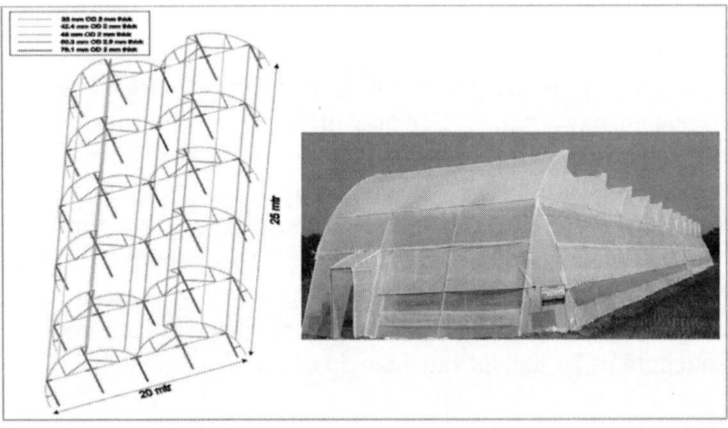

Fig. 1: Saw-tooth greenhouse

6. Quonset greenhouse

Support is provided by pipe arches that span the length of the quonset greenhouse. Polyethylene is commonly used as a covering material. When compared to gutter-connected greenhouses, this greenhouse is ideal for a small isolated cultural area and is also cost-effective. There are either freestanding dwellings or ridge-and-furrow dwellings. A plant bed can develop between the overlapping parts of neighbouring dwellings when the truss elements are interlocked.

C. Type of greenhouse based on utility

The utilities and purposes of a greenhouse can be used to classify it. For example, artificial cooling and heating are more expensive and complex. By using artificial cooling and heating, greenhouses are classified as active heating and cooling systems.

1. Active heating system of greenhouses

At night, the temperature inside the greenhouse drops. To keep plants from getting a chilly bite from freezing, some heat must be provided. The amount of energy required to heat a greenhouse is determined by the pace at which heat is lost to the outside environment. In order to reduce heat losses, different technologies are used, such as double-layer polyethylene, thermopane windows (two layers of factory-sealed glass separated by a dead air gap), and heating systems such as unit heaters, central heating, radiant heating, and solar heating systems.

2. Active cooling system of greenhouses

For optimal crop development throughout the summer, it is preferable to keep greenhouse temperatures lower than ambient temperatures. As a result, changes to the green house are made to allow huge quantities of cooled air to be pulled into the greenhouse; this form of greenhouse uses either an evaporative cooling pad with fan or fog cooling. With this greenhouse, you can open the roof up to 40 %, and in some cases even 100 %.

D. Types of greenhouse based on construction

The structural material dominates greenhouse construction, while the covering material has an impact on the different types of greenhouses. The length of the green house affects the structural components that will be used and how they will be built. The larger the span, the more efficient the material should be, and more structural components should be utilized to create robust truss type frameworks. Simpler designs, such as hoops, can be used for shorter spans.

Greenhouses are divided into three types based on their construction: wooden framed, pipe framed, and truss framed buildings.

1. Wooden framed structures

In general, timber frame constructions are appropriate for greenhouses with spans under 6 meters. Without the use of truss materials, the side posts and columns are built of wood. Pine wood is often utilized since it is both cheap and strong. Similarly, locally available lumber may be utilized for greenhouse construction since it is strong, durable, and machinable.

2. Pipe framed structures

When the clear span is approximately 12 meters, pipes are utilized to create greenhouses. Pipes are used to create side posts, columns, cross ties, and purlins in general. The trusses aren't employed in this design.

3. Truss framed structures

Green house structures can be built in truss frames if the span is at least 15 meters. A truss is made up of rafters, chords, and struts that are welded together from flat steel, tubular steel, or angular iron. Struts and chords are compression and tension support members, respectively. Throughout the greenhouse, angle iron purlins connect each truss. Columns are typically used only in very broad truss frame homes of 21.3 meters or more. The majority of glass homes are constructed using truss frames, which are ideal for pre-fabrication.

E. Types of greenhouse based on covering materials

A greenhouse's structure is most important in terms of its covering materials. They are directly impacted by the greenhouse effect and regulate the air temperature inside the greenhouse. The covering materials influence the frame type and manner of installation. Glass, plastic film, and rigid panel greenhouses can be classified according to their covering material.

1. Glass greenhouses

Glass is used as the covering material in glass greenhouses (Fig. 2). Its benefits include greater interior light intensity, greater air infiltration, lower interior humidity, and superior disease prevention capabilities as a covering material. Glass greenhouses are constructed using ridge and furrow, lean-to, and even span designs.

Fig. 2: Glass greenhouse

2. Plastic film greenhouses

Flexible plastic sheets, such as polyvinyl chloride, polyethylene, and polyester, are used to cover this sort of greenhouse. Plastics are more common as a greenhouse covering material since they are less expensive and require less heating than glass greenhouses. The primary downside of using plastic sheets as a covering material is that they have a short lifetime. Even the finest ultraviolet (UV) stabilized film only lasts four years. This covering material is appropriate for use in both quonset and gutter-connected designs.

3. Rigid panel greenhouses

In greenhouse construction, quonset type frames consist of rigid polyvinyl chloride panels, fibre glass-reinforced plastic panels, acrylic resin panels, and rigid polycarbonate panels. These materials, unlike plastic or glass, give a more consistent light intensity throughout the greenhouse and are also more resistant to breaking. High-quality panels have a longer lifespan, up to 20 years. However, the panels move, collecting dust and harboring algae, which is the major drawback of rigid panel greenhouses, resulting in dimming of the panels and a progressive decrease in light transmission.

Structural components for greenhouse construction

Rafters are the major vertical supporter of greenhouses in terms of structural components (Fig. 3). Rafters are often spaced at 2, 3, or 4 feet intervals, depending on the amount of energy required to support the entire building. Depending on the breadth of the greenhouse, it might be truss, curved, or arch. Purlins go from rafter to rafter and offer horizontal stability. Based on the size of the greenhouse, all structural components are placed 4-8 feet apart. Purlins, linked by a cross tie, are used in high-wind areas to offer additional support. Vertical support is provided by side posts and columns, which typically range in height from 1 to 10 feet. The height of the manufacturing area is determined by these structural components. Cooling is provided by sidewalls, while insulation is provided by adequate ventilation.

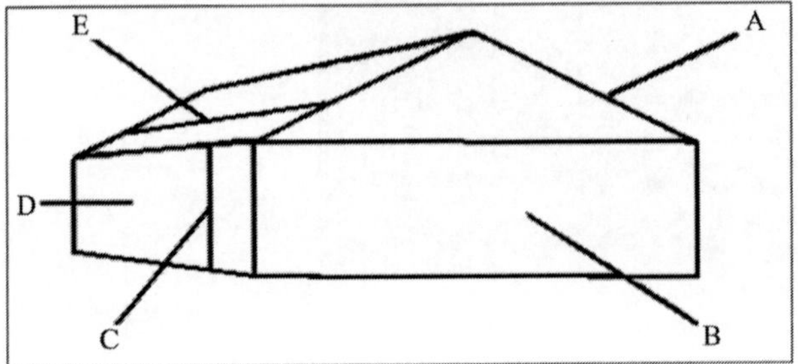

A. Rafter, **B.** End wall, **C.** Side post, **D.** Side wall and **E.** Purlin
Fig. 3: Primary components of greenhouse

The design of a greenhouse has a significant impact on its productivity and efficiency. Ridge and furrow greenhouse design provides significant productivity and efficiency in output, and aluminium is the most dependable and widely utilized frame material for trade greenhouse structures. Similarly, for crop production, multiple sheets of polyethylene film are the most cost-effective covering material. However, the initial facilities and long-term expenses, the lack of availability of various structural components, the lack of standardization of region-based greenhouse and other structure design, and a lack of knowledge are all key barriers to its implementation.

2. Hot frames (hotbeds) and heated sun tunnels

A hotbed (also known as a hot frame) is a tiny, low structure that performs many of the same tasks as a propagation house (Fig. 4). A hotbed is typically a wooden frame or box with a slanted, snug-fitting cover made from window sashes. Except for places with harsh winters, where they can only be used in the spring, summer, and fall, hotbeds can be used throughout the year. Other types of hotbeds can be made out of hooped metal tubing or bent PVC pipe coated with polyethylene to maintain a constant temperature (sometimes a white poly material is used to avoid the higher temperature build up and temperature fluctuations of clear poly).

The size of the frame is traditionally determined by the size of the glass sash available a common size is 0.9 by 1.8 m. (3 by 6 feet). Any appropriate size can be utilized if polyethylene is used as the covering. The frame is simply constructed by nailing 3 cm (1 inch) or 6 cm (2 inch) timber to 4 × 4 corner posts. As much as possible, use decay-resistant wood, such as redwood, cypress, or cedar. Pressure-treat wood with preservatives, such as chromated copper arsenate (CCA). This chemical slows decomposition for a long time and does not emit plant toxic gases. Creosote should not be put on wood buildings where plants will be grown since the gases emitted are harmful to plants, especially on hot days.

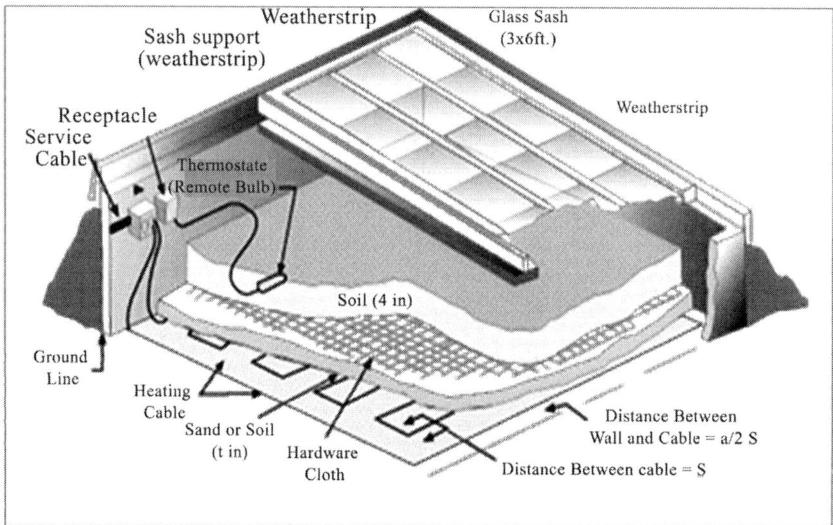

Fig. 4: Construction of an electrically heated hotbed

Plastic or PVC tubing with recirculating hot water is ideal for providing bottom heat in hotbeds. Over the hot water tubing, a rooting or seed germinating media of 10 to 15 cm (4 to 6 inch) is poured into the hotbed. Alternatively, flats with liner pots containing the medium or community propagation flats can be utilized. These are placed immediately on the hot water tubing, which is covered with a thin coating of sand.

Early in the season, seedlings and leafy cuttings can be started and rooted in hotbeds. Shade, ventilation, as well as temperature and humidity must be carefully managed, just like in a greenhouse. Hotbed structures are excellent for generating many thousands of nursery plants in small propagation operations without the greater building costs of bigger, walk in propagation facilities (Hildebrandt, 1987).

3. Cold frames and unheated sun tunnels

One of the most common uses of cold frames is to condition or harden rooted cuttings or seedlings before planting them in the field, nursery row, or container. Plants can be started in late spring, summer, or autumn without having to provide external heat (Wood, 1985) with cold frames and unheated sun tunnels. In addition to poly-covered hoop houses, cold frames include poly-covered wood frames or unheated sun tunnels where people cannot go through. To retain heat and achieve high humidity, the covered frames should fit firmly. The sash cover of cold frames slopes down from north to south, so they should be placed in wind-sheltered areas (south to north in the Southern Hemisphere).

The construction of a low-cost cold frame (Fig. 5) is similar to that of a hotbed, with the exception that no provision is provided for providing bottom heat. When covering older-style cold frames, a lath covering with wide gaps between the lath boards is occasionally utilized. Despite this, freezing temperatures are not prevented, but temperature swings between high and low are minimized.

Fig. 5: Low-cost cold frame

Only the heat of the sun is used in these buildings, which is kept by the transparent or opaque white polyethylene covers. Winter protection, shade, ventilation, and watering are vital for a cold frame's success. At first, when the young, tender plants are placed in a cold frame, the covers are kept tightly closed to maintain a high humidity level. However, as the plants become acclimated, the sash frames are gradually raised or the ends of the hoop house or sun tunnels are gradually opened to allow for more ventilation and drier conditions.

Maintaining humid conditions in a cold frame requires the installation of a mist line or regular watering of plants. Unless ventilation and shade are provided, temperatures in closed frames can rise to dangerously high levels on hot days. The sash can be covered with spaced lath, Saran or poly shade cloth, or reed mats to offer sun protection. During extreme cold weather, plants overwintered in cold frames may require additional protection.

4. Lath houses

Shade houses (Fig. 6) provide outdoor shade, which shields container-grown plants from high temperatures and strong light irradiance during the summer (Gordon, 1988). Plants' water requirements are reduced and moisture stress is reduced. Among the many uses of lath houses in propagation, hardening off and acclimatizing liner plants before transplantation, as well as caring for shade-loving or sensitive plants, are just a few examples. Nurseries will occasionally utilize a lath house to store plants for sale. In warm regions, they are used for propagation, as well as a misting facility, and can even serve as overwintering structures for liner plants.

In higher latitude locations, snow load might be a concern. The construction of a lath house varies greatly. There are prefabricated aluminium lath houses available; however, they may be more expensive than wooden buildings. Most often, pipe or wood supports are placed in concrete with cross-members supporting them. The majority of lath houses are now covered with high-density plastics such as polypropylene fabric and UV-treated polyethylene shade cloth, which come in different shade percentages and colors. Various densities of these materials allow plants to receive less light, such as 50 % sunshine. They are lightweight and may be connected to supporting poles with strong wire. The shade fabric is rip-resistant and has a lifespan of 10 to 15 years, depending on climate and material quality. Producers will wrap the shade cloth with polyethylene for winterization in colder climates. Some shade systems use 5 cm (2 inch) wide strips of wood, which can cover one-third to two-thirds of the area. Typically, all sides and the top are covered. Snow fencing rolls connected to a supporting structure can be used to build a fence at a low cost.

Fig. 6: Lath houses or shade houses

5. Propagation frame

Even in a greenhouse, humidity levels aren't usually high enough to allow some types of leafy cuttings to root successfully. Poly or glass enclosures may be needed for successful rooting. Such gadgets come in a variety of shapes and sizes. In the past, little ones were referred to as Wardian instances. These enclosed frames are excellent for graft union formation in tiny potted nursery stock because they preserve high humidity.

When taking fall semi-hardwood cuttings in chilly summer regions, a sheet of extremely thin (1 or 2 mils) polyethylene put immediately on top of a bed of freshly made leafy cuttings in a greenhouse or lath house can sometimes provide enough relative humidity to ensure successful roots. A contact polyethylene system is another name for this. This method requires good shade management to minimize light irradiation.

If you want to speed up the development of graft unions on a smaller scale, bell jars (big inverted glass jars) can be placed above unrooted cuttings or recently grafted containerized plants. In such devices, humidity is maintained at a high level, although temperature regulation requires considerable shading. All of these structures must be used with caution to avoid the accumulation of harmful organisms. Temperatures are too warm and humid, as well as a lack of air circulation, leading to the growth of bacteria and fungi. The cleanliness of all items placed in such units is critical; yet, fungicides may be required on occasion.

6. Net house

In tropical locations where artificial heating is not necessary and artificial cooling is costly, net homes are commonly utilized as propagation structures. In these regions, net homes with glass or plastic film tops and wire net sides may be erected. It offers adequate ventilation and maintains an optimal temperature for seed germination and subsequent seedling growth. To shut off solar radiation energy and keep the dwelling cool, the roof of the net house can be covered with gunny fabric or even live plant creeper. Net houses may be built to meet the needs of the propagator, thus their size varies according to the nurserymen's needs. Figure 7 depicts a standard shade net.

Fig. 7: Net house

7. Bottom heat box

It is a basic box for encouraging cutting roots in hard-to-root fruit plants like mango and guava. It is made up of two galvanized iron sheets chambers. The outer chamber is 70 cm in height and 46 cm in width, while the inner chamber measures 68 cm in height and 44 cm in width. For heat insulation, the gap between the two chambers is filled with glass wool.

Another chamber with a height of 35 cm is installed within the inner chamber, with two electric lamps at the bottom to provide heat to the cutting. The cuttings are placed into the innermost chamber, which is filled with soil mixture or any other propagation media. The cuttings are heated and illuminated by two 100-watt electric lamps installed at the bottom of the chamber. Similarly, a thermostat installed at the bottom of the chamber maintains and controls the temperature in the box automatically. The best temperature to keep in the box is 30 ± 2°C because cuttings of mango, walnut, olive, and guava root quickly and abundantly at this temperature. The time it takes for a cutting to begin rooting varies per species, but in general, it takes 1-2 months for the roots to grow properly.

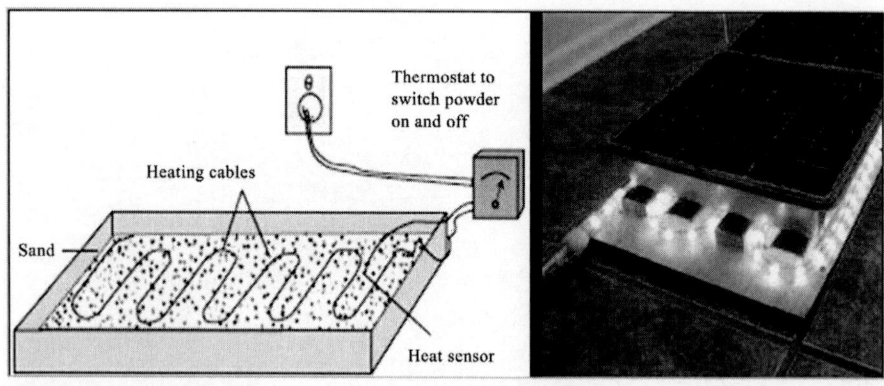

Fig. 8: Bottom heat propagation box

8. Mist propagation unit

The spraying or misting of softwood leafy cuttings to root them is a technique that nurserymen and other plant propagators employ all around the world. The goal of misting is to keep a coating of water on the leaves at all times, decreasing transpiration and maintaining the cuttings turgid until they root. Despite the high humidity, leafy cuttings can be fully exposed to light and air in this manner, preventing damage even from direct sunlight. Mist also protects cuttings against disease infection by washing away fungal spores that assault the tissues. While the leaves in this procedure must be maintained wet at all times, it is critical that just a small amount of water be utilized. This is due to the fact that too much water leaches nutrients from the compost, causing hunger in the developing plants. Furthermore, overwatering might have a direct negative impact on the clippings. As a result, nozzles capable of generating a very thin mist must be used. Small mist propagation units are generally employed by small farmers (Fig. 9), but commercial nurserymen in advanced nations utilize more advanced intermittent equipment.

In most modern nations, mist propagation machines are utilized to propagate difficult-to-root cuttings. A greenhouse is used to build mist beds. Throughout the day and night, a fine mist is sprayed occasionally over the clippings. A time clock controls the mist unit, which operates a magnetic solenoid valve and is configured to turn on. Turn on the mist for about 3-5 seconds to wet the leaves, then turn it off for a while and turn it back on again when the leaves are dry.

The mist has five control mechanisms in total. Screen balance and photoelectric cell, as well as a timer, electronic leaf, thermostat, and timer. Generally, a mist unit has two types of timers: one that runs in the morning and shuts off at night, and another that runs during the day to produce an intermittent mist, generally 6 seconds on and 90 seconds off. Under the mist with cuts in the leaf, a plastic with two terminals is positioned, and the alternation between drying and wetting shuts off the current, which turns off the solenoid valve. The temperature of the mist is controlled by a thermostat. A stainless-steel screen is connected to a lever with a mercury switch in the screen balance control mechanism. When the mist is turned on, water collects on the screen, and when the weight of the water exceeds a certain threshold, the mercury switch is tripped. A photoelectric control is based on the connection between transpiration rate and light intensity.

Fig. 9: Mist Propagation Unit

The benefits of using mist units

This method allows for the rooting of difficult-to-root plant cuttings. It allows for the easy usage of soft, succulent material throughout the growing season, which is far more likely to root than older, mature, and hardened wood. It keeps slow-rooting cuttings alive for a longer length of time, allowing them to root before succumbing to desiccation. Using this method. Large cuttings with a lot of leaf area can be rooted, but this restricts the number of large, salable plants that can be produced in a short period of time.

In a glasshouse or a polyethylene tunnel, a mist unit can be set up. Typically, propagation beds with a width of 1.2 m are built up. The configuration of the jets is critical. A mist unit may be set up by installing a mist propagation unit, however all of the jets must be at the same height. A sufficient quantity of water is required for the mist to work properly. The water should be free of salts and have a good pressure. If using alkaline water, it is best to avoid clogging the nozzles in the mist chamber and forming deposits on cuttings' early leaves, which will hinder root growth and development. Furthermore, a well-drained rooting medium must be utilized, and there must be enough provision for the drainage of excess water. Furthermore, blue and green algal growth is quite frequent in mist propagation structures, which is very damaging to the propagating material, thus every precaution should be made to maintain the mist propagation unit free of algae. After cuttings have been rooted in the mist, they must be hardened in order to be successful in the field. Misting should not be stopped suddenly after cuttings have been rooted, since this may cause the young plants to dry out and burn. As a result, a weaving off procedure should be used, in which misting continues but the number of sprays/days progressively decreases. It may be accomplished by reducing the 'on' times and increasing the 'off' ones. Another option is to move the rooted cuttings to a greenhouse, fog chamber, and frames where they will be kept at a warmer temperature and with low relative humidity. The rooted cuttings are only planted at permanent places after phase wise hardening.

9. Plastic tunnels

Greenhouse Crops Research Institute, Littlehampton, England, devised a simple but efficient way of preserving cuttings that is now used by many nurserymen across the world. It comprises of a plastic tunnel similar to those used to preserve strawberries. The 0.2 inch diameter wire hoops that support the polyethylene are built at 30 inch intervals and are three feet wide. The hoops are then covered with a white translucent polyethylene sheet 6 feet wide, which is kept in place by connecting two lines of polyethylene bailer wire to alternate hoops on either side of the tunnel. During the winter, seeds, cuttings, and other propagating material are sown/planted inside the tunnels (Fig. 10). Perforated tunnels are now utilized instead of non-perforated tunnels. Temperatures and relative humidity are rapidly rising, potentially affecting cutting roots and propagating material growth.

Fig. 10: Low-cost plastic tunnel

10. Growing rooms

A growth chamber is an enclosed structure that is generally devoid of natural light. Artificial lighting is used to provide illumination. In most sophisticated nations, growing rooms are now frequently utilized professionally for the development of seedlings of bedding plants, tomatoes, and cucumbers. Typically, seedlings are cultivated in trays or pots on benches. The seats are generally built in layers that are vertically approximately 2 feet long and 6 inches broad to conserve space. Eight feet long, 125 watt fluorescent bulbs are placed 1 feet 6 inches above each bench to provide illumination. Each bench has seven tubes that produce 500 lumens per square foot of light, which is plenty for bedding plants. More fluorescent tubes can be put over the benches housing the plants if the plants produced in such dwellings have higher light needs. The tubes' heat maintains a temperature of at least 70 degrees Fahrenheit, and extreme temperatures are avoided by using fans installed in the building.

11. The automated greenhouse

Modern greenhouses may now be nearly entirely automated, making propagation easier. The temperature of the air and the bed, for example, may be regulated as needed with the use of thermostats. Automatic ventilation, on the other hand, permits ventilators to open and close in response to temperature. Even in modern greenhouses, automatic irrigation systems are built, and water is delivered to the plants *via* a drip or trickle system to each pot or plant via a timed nozzle. Automatic 'on' and 'off' valves or a time switch are used to regulate this system.

The utilization of phytotron facilities has made the most progress in growing plants in highly regulated environments. Scientists are utilizing the National Phytotron Facility at IARI in New Delhi to grow plants by adjusting the temperature and relative humidity to their specifications.

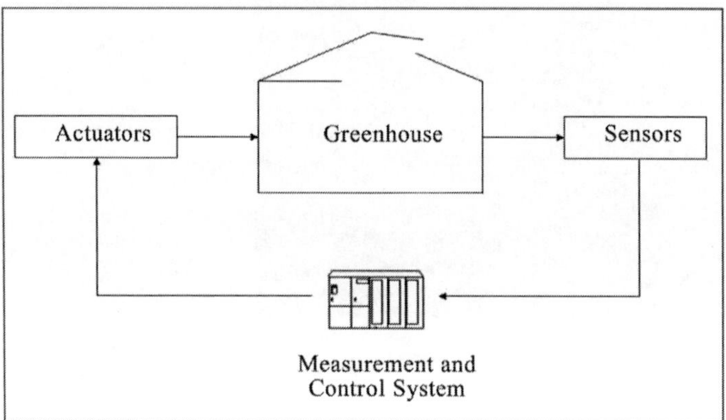

Fig. 11: Shows a software to automate greenhouses

Structures and their possibilities

Structures are utilized to create an environment that can be readily controlled and maintained. If a structure loses this characteristic, it becomes less valuable to the propagator, which might result in financial losses. Control device failures, such as probe malfunction, incorrect interpretation of information conveyed by the instrument, or the structural frame itself, are all possibilities. Links and joints that hold propagation structures together might weaken with time. As a result, the structure's connections, joints, and overall stability must be examined on a regular basis. Cracks or rips in the covering material, whether it's polyethylene sheeting or shade cloth, cause unwelcome changes in the structure's environmental conditions. Plants are exposed to harsher circumstances, which increases the likelihood of plant damage.

Greenhouses are more vulnerable. Failures of fans and moist walls might have severe consequences. Temperatures can quickly rise to the point that plants are unable to carry on with their regular metabolic activities, resulting in severely wilted plants due to a poor photosynthesis rate. A lack of aeration can cause a decrease in CO_2, inhibiting photosynthesis. With a continuous supply of O_2 and CO_2, high respiration rates are necessary. The incorrect shade house covering material might result in a reduction in light intensity at certain times of the year. Before reaching the plants, light is filtered. The amount of light received is influenced by several factors, including the location and season.

From October through March, the light intensity is at its highest, peaking in January. The light intensity is moderate from April to September, with the lowest reported in June (21st).

References

Anonymous, 2020. Food and Agriculture Data, FAOSTAT, Rome.

Beytes, C. 2003. The Ball Red Book: Crop Production. 17th ed. Vol. 2. Chicago, IL: Ball Pub.

Dalai, S., Tripathy, B., Mohanta, S., Sahu, B. and Palai, J. B. 2020. Green-houses: Types and Structural Components. In: Protected Cultivation and Smart Agriculture edited by Maitra, S., Gaikwad D. J. and Shankar, T. © New Delhi Publishers, New Delhi, pp. 09-17.

Gordon, I. 1988. Structures used in Australia for plant propagation. *Proceedings of the International Plant Propagator's Society*, 37:482–89.

Hildebrandt, C. A. 1987. Economical propagation structures for the small grower. *Proceedings of the International Plant Propagator's Society*, 36:506–10.

Wood, J. S. 1985. Sun frame propagation. *Proceedings of the International Plant Propagator's Society,* 34:306–11.

9

Nursery Establishment

R.K. Jat, S.K. Acharya and Mukesh Kumar

A nursery is a place, where seedling, saplings, trees, shrubs and other plant materials are raised and sold out for planting in gardens and orchards. The prerequisites of a successful and remunerative fruit production are the availability of true-type, healthy and good quality planting materials. Nursery is, therefore, a place where one can produce true to type plants by gaining technical skill, maintaining plants property and by careful planning. Like orchard establishment, establishment of a nursery is also at permanent venture and any mistake made initially can't be rectified easily in the later stages. Therefore, for establishing a nursery, due care in selection of site, plants to be raised and transportation facilitation *etc.* must be given.

Pre-requisites for establishment of a nursery

When establishing a nursery, keep the following points in mind

1. The nursery should be situated as close as possible to the important production areas.
2. The soil should be deep, fertile, well-drained and free from soil borne pathogens.
3. The locality should have adequate supply of sweet water.
4. A favourable climate is needed to propagate the plants.
5. The site should be easily accessible and connected by various means of communication.
6. To handle the different operations, there should be sufficient labour, bulldozers, and grafters available.
7. There should be a ready supply of materials such as fertilizers, pesticides, growth regulators, grafting wax, lanolin paste, and equipment.
8. It should be possible to build different propagation structures, such as glasshouses, lath houses, *etc.*

9. The nursery should have its own resources for providing mother plants (parent plants) for propagation.

Importance of nursery

1. It is important to pay special attention to young seedlings during the first few weeks after germination. Nursery beds are easier to maintain in a small space than large permanent sites for the young and tender seedlings.
2. Vegetative propagation is the most common method of propagation of fruit crops. Propagules need special care before being transferred to the main field. During a controlled nursery environment, skilled labour can provide all these tasks successfully.
3. The mist house chamber, which is integral to a nursery, is the best place to root cuttings and harden grafts.
4. The direct sowing method did not perform well in several crops when compared to transplanting in nursery-grown seedlings.
5. Replacement of causalities in orchards is best handled by plants hardened in the nursery.
6. Furthermore, raising seedlings or saplings in the nursery gives farmers more time to prepare for planting.
7. Seasoning/hardening of seedlings against natural odds is only possible in nursery.

Classification of nursery

On the basis of their site, nurseries can generally be divided into two groups:

1. Home nursery 2. Commercial nursery

1. Home nursery

An area where plants are cultivated or raised exclusively for the garden of the grower. The area is small and the primary consideration is the raising of quality planting material. Therefore, expensive methods of nursery practices are adopted.

2. Commercial nursery

A nursery has a larger collection of plants and a larger size. This is primarily concerned with the economic return from investments, and very expensive methods are avoided, it may be a government or private nursery. This type of nursery can be divided into two groups which are as follows:

(a) **Rural nursery:** This type of nursery is located in a village near a high way or a railway station. Rural nurseries are usually larger than urban nurseries because labour and land are more affordable. The planting material in such nurseries is sold out at a cheaper rate because the cost of raising the planting material as well as the buying power of the indentors is less.

(b) **Urban nursery:** Nursery of this type is located in a city or town. A nursery is usually small because the land is expensive and not easily available. The labour charges, transport cost *etc.,* are also very high, but these are compensated by the higher price of products and volume of sale. Sometimes they act as middlemen *i.e.*, procure planting materials from rural nursery and resale to the customers.

The nursery can be grouped into different types based on the business which are wholesale nursery, the retail nursery, the landscape nursery, the agency nursery and the mail order nursery.

1. Whole sale nursery

Typically, plants are produced in large quantities for retail sale in this kind of nursery. In general, these nurseries are loved in rural areas, where the land and labour are available at cheaper rates. Due to these reasons, rural nurseries can afford to expand their business without incurring too much additional expenses.

2. Retail nursery

As retail nurseries depend mostly on house owners for their trade, they purchase plants from whole sale nurseries. They must be located in or near a city. Besides plants, these nurseries also provide items such as fertilizer, seeds, and tools.

3. Landscape nursery

The landscape nursery should be situated near a populated area since urban people require landscape plants to beautify their homes. Such nurseries should be located on the out skirts a city or town as it facilitates the clientele to obtain landscape service.

4. Mail order nursery

This is a specialized whole sale nursery. Mostly, it sells stock based on a catalogue. Users order from the catalogue and receive the plants via mail or courier. These nurseries are also located in a locality, where land is comparatively cheap and sufficient labour, water and transportation facilities are available.

5. Agency nursery

A nursery agency sells its stock through agents or sales representatives. They are usually highly specialized and are few in number.

Factors affecting the establishment of a nursery

1. Location and site

i. **Topography:** The land for the nursery should be as flat as possible with little slope in one direction. This helps in quick draining out the excess water during the heavy rain. The area should be slightly elevated than the adjoining area to avoid water logging and congestion.

ii. **Climate:** It is of prime importance to select only those species and varieties which thrive well in the region.

iii. **Reputation of locality for business:** The area selected for nursery establishment should be reputed for one or the other kind of business. It should be a known locality. Since such a place is known to the customers, the fledgling nursery gets the advantage of publicity. In addition to this, skilled labours and various inputs are also readily available.

iv. **Transport facility:** Adequate transport facilities are of prime importance for successful nursery business. The area selected for nursery establishment should be well connected with road or by a rail track to deliver the plant materials to the customers in a short time. It should be easily approachable by the customer.

2. Selection of soil

Soil type, drainage and soil fertility are the three basic components to be considered while selecting the soil for establishing a nursery. Almost all the fruit crops do well in friable, loamy soil, rich in organic matter and having a pH range of 5.5 to 6.5. However, a clay-loam soil is advantageous to lift the evergreen plants easily with earth balls intact. Shallow soil with hard and compact subsoil should be avoided as it creates hindrance in proper root growth of mother-plants. Usually, a depth of 70-80 cm is sufficient for growing plants in nursery but for the progeny block a soil depth of 1-1.5 m is desired. Presence of calcium carbonate layers affects permeability and aeration, and results in stagnation of water which is detrimental to root respiration.

3. Water facility

Nursery stocks require frequent light irrigation. A regular water supply must be assured in a nursery. The water for irrigation should be free from any dissolved salts and pH should be near neutral. A surface well or a tube well may be dug for this purpose. In canal-irrigated areas, it is advisable to construct a water reservoir, so that the requirements may be met with the stored water during the dry period.

4. Manures

For nursery work, plenty of organic manures like FYM, compost, leaf mould, compost *etc.* are required. These must be available in the locality in sufficient quantities.

5. Availability of labour

Sufficient labours are required for running of nursery activities easily. These should be available locally at reasonable rate.

Components of nursery

The following components should be included in a nursery:

1. Building structures

This includes offices, sale counters, packing sheds, potting sheds, stores, implement sheds, bullock sheds, *etc.*

Office: The office is the most important amongst all the building structures because all the business transactions are made from this office. It should look beautiful and attractive for customers. The size of the office depends on the size of the nursery. The store and office can be in the same building.

Sale counter: The sale counter should be attached to the office.

Packing shed: Before delivering the material to the customers they are to be packed and labelled properly. Packing materials *viz.*, baskets or boxes are kept in this shed. The shed should be located near the sale counter.

Potting shed: Pots of varying sizes and shapes and the potting mixture are kept in the shed. Potting operations are performed under shade where all the materials are easily available in perfect condition throughout the year. Only the plants are brought from the nursery beds. Potting under shade also minimizes desiccation of plant and saves time.

Store: A store is needed to keep the fertilizers, sprayers, insecticides, fungicides and planting materials *etc.* This should be located inside the nursery and away from the main office.

Implement shed: A shed is required for keeping the garden implements.

Bullock shed: Many bullock drawn implements are generally used in a small nursery. In such case, a shed is required and should be located in the rear side of the nursery to maintain the bullock heard of the nursery.

Residential quarters: Sometimes nursery holder desires to live in the nursery premises with his family. In such case, the house should be constructed as per choice of the owner. In large enterprises, residential provisions are also made for the skilled labours and other staff members.

2. Progeny tree block

The selection of fruit crops and the collection of true to type mother plants have a strong bearing on the success and quality of a nursery industry. A suitable fruit crop should be chosen to meet customer demands. To provide a wide choice, a number of promising varieties of popular crops should be collected. The progeny tree should be healthy, disease-free, genetically true to type, and pest-free. The pedigree of these plants should be known to the nursery man.

3. Propagation structures

Several times, the outdoor conditions are not favourable to raise plants successfully. Low or high temperature or the hot desiccating winds may be harmful for the seedlings and their subsequent growth. Therefore, modern propagation structures are needed, such as greenhouses, hotbeds, cold frames, lath houses, net houses, mist chambers, *etc.* to create favourable conditions in respect of light, temperature and humidity for facilitating germination of seed, rooting of cutting and also for hardening of young seedlings before transplanting them in the field. The green house and hotbed are mainly used for seed germination and rooting of cutting owing to arrangements of temperature control and plenty light. A lath house or cold frame is a structure used for hardening young and tender plantlets before they are planted outdoors.

4. Nursery beds

A nursery bed should be prepared by repeated ploughing and pulverising the soils to obtain a fine fifth. Appropriate amount of organic manures and fertilizers should be incorporated into the soils. A loose, friable, clay loam soil, rich in organic matter is ideal for the young tender seedling. There must be an adequate provision for proper drainage because the young seedlings are sensitive to water logging condition. Usually, nursery beds have 3 meter length and 1 meter width. Nursery beds should be little raised during rainy season to avoid stagnation of water.

5. Packing yard and working shed

The packing yard is used to pack the plants for sale to the outstations. It can be combined with the working shed. As a result, there should be ample space for a number of workers to sort out and pack the plants with ease.

6. Pot yard

Pot yards should be shaded since tender plants need shade more than hardy ones. The plants in this section will need to be watered frequently, so it should be near a water source. The multiplication of plants can be accomplished using plastic bags, root trainers, earthen trays or pots of the right size. Plastic pots and poly-bags are convenient and easy to handle.

7. Compost pit

It is prepared to decompose waste to get it decomposed as compost. Compost is used as a media for growing seedlings and raising cuttings. It is also used as a component in pot filling mixture. It is an ugly looking site. It is advisable to have provision for it in one corner of the nursery under the shade of the tree. Compost or any media can be sterilized with 2 % formation solution during warmer period. Various media are also used in the nursery *i.e.* soil, sand, sphagnum moss, perlite, vermiculite, peat moss and saw dust *etc.*

References

Chattopadhyay, T. K. 2012. A Text Book on Pomology. Volume 1: Fundamentals. 5th Edition, Kalyani Publishers, New Delhi.

Sharma, R. R. 2002. Propagation of Horticultural Crops: Principles and Practices. Kalyani Publishers, New Delhi.

Singh, J. 2014. Basic Horticulture. Kalyani Publishers, New Delhi.

10

Media for Propagation

Ajit Kumar Singh and Ashok Dhakad

Nurserymen and propagators propagate different types of horticulture plants with seeds, cuttings, and other methods. Regardless of the medium used, an ideal medium should have the following characteristics:

1. The medium should be sufficient firm and dense to told seed or cutting or layer firmly in place during germination or rooting.
2. The volume of the medium should remain constant upon drying or wetting.
3. This medium should be able to hold and deliver an adequate amount of moisture to the seeds, cuttings or layers.
4. This material should be porous enough to allow adequate aeration and to drain away expensive moisture.
5. The medium must be free from seeds of unwanted plants (weeds) and pathogens, which may be the germination of seeds or rooting of cutting, layers *etc.*
6. It must be free from high levels of salts.
7. Its pH should be between 5.5 to 6.5.
8. It should be able to provide adequate nutrition to the germination seeds or cutting or other plant parts.
9. The pasteurization process must be able to be performed with steam or chemical without causing harm.
10. It should be cheap and easily available.

Criteria for the selection of a medium

The following criteria should be taken into consideration while selecting the medium.

1. Whenever possible, we should choose locally available materials.
2. Check the quality of the medium before buying it.
3. Take guidance of a technical expert or a person dealing in nursery business.
4. The selection medium should have sufficient and ideal pH (5.5 to 6.5).
5. The selected medium should easily mix with other growing medium.

Commonly used media

The characteristics of commonly used media by propagators and nurserymen are described below:

1. Soil

Soil is the part of earth's crust that is formed by the decomposition of rocks and minerals under the influence of physical, chemical, and biological forces. Soil consists primarily of solid, liquid, and gaseous states. For proper growth of the plants all these contents should be in a proper proportion. Solid portion of soils has both organic and inorganic matters which determine the structure and texture of the soil. The inorganic part is mostly the residue, which is the resultant of decomposition of parent rocks by weathering process. These vary in size, ranging from large sized gravels to extremely minutes particles of clay. The coarse particles serve as supporting framework of the soil and the clay fraction the store house of the nutrients which are absorbed by the plants. The organic portion of the soil consists of living and dead organisms (fungi, bacteria, and insects). Humus is composed of decayed organic matter and dead organisms. It is highly colloidal in nature and is responsible for storing water and plant nutrients. Soil is mainly comprised of water containing dissolved mineral elements, dissolved oxygen, and dissolved CO_2. The gaseous portion of soil is very important for plant growth in waterlogged (poorly drainage) soil in which water replaces the air, depriving plant roots of essential microorganisms that are vital for proper growth and development of plants.

A soil's texture is an important physical property. This determines the success or failure of seed or cutting or other propagule. It depends on the relative proportions of sand, silt, and clay. The main textural classes of soil include sand, sandy loam, silt loam, clay loam and clay *etc.* A soil containing 40 per cent sand, 40 per cent silt, and 20 per cent clay is considered the best texture class for germination by many plant species. Sandy loam soil is excellent for preparing soil mixtures for container gardening. The pH of soil also determines the success or failure of seed germination or rooting of cutting or layer. In general soil having neutral pH (5.5 to 6.5) is always preferred for better germination of seeds and rooting of cutting, or layers. In addition to texture, the structure of soil, which refers to the arrangement of soil particles within a soil mass, plays an important role in the germination of seeds and the structure of cuttings or layers. Water holding capacity of soil is usually determined by its structure. Thus, a soil having good texture with water holding capacity should be preferred for using as a propagating medium. In general, heavy soils should be avoided, as working with these soils is very difficult. Moreover, these soil from heavy clouds have very high water holding capacity which is injurious both for seed germination and rooting cuttings or layers propagules.

2. Sand

Sand is formed by weathering of rocks. Its composition varies according to the type of rock formed. The size of sand varies from 0.05 to 2 mm in diameter. Sand contain silica with almost no minerals quartz sand, which consist chiefly compounds, is mainly used for propagation by the nurserymen. Sand is the heaviest of all rooting media, so it should be used in combination with some other organic material. Commercially, sand is available in many grade and sizes. However, sand should preferably be washed, fumigated or heat treated before use to kill harmful pathogens present in it.

3. Vermiculite

Vermiculite is a micaceous mineral that is widely used as a propagation medium in advanced countries. It is a hydrated aluminum iron silicate that expands rapidly when heated. It becomes very light when it expands. The substance is insoluble in water, but it absorbs a large amount of water. It has a high capacity for cation exchange and has the ability to hold nutrients for longer periods of time. There is adequate Mg and K to support plants' growth.

Horticultural vermiculite has four grades, depending upon the size of the particles; No.1 grade has particles from 5 to 8 mm in diameter, No.2 from 2 to 3 mm, No.3 from 1 to 2 mm and No.4 from 0.75 to 1mm. In general, No. 2 grade vermiculite is a regular widely used horticultural grade and is widely used. No. 4 grade vermiculite is best used as a seed germinating medium.

4. Peat

In a partially decomposed state, it consists of the remains of aquatic flora from marshes, swamps, or bogs that have been preserved under water. It is derived from sphagnum, hypnum or other mosses. It is called peat moss. It is mixed with water after breaking and moistening.

5. Sphagnum moss

Commercial sphagnum moss consists of the dehydrated residue or living portion of acid bog plants like *Sphagnum papilliosum*, *S. capillacem* and *S. palustre*. During the rainy season, it is generally collected from the trunks of trees in south Indian hills above 1500 m above the mean sea level. It is relatively sterile, light in weight, and has a high capacity for holding water. It is the most commonly used air layering medium.

Container for propagating and growing young plants

1. Earthen pots

The pots are made of burnt porous clay in various sizes to provide appropriate amounts of soil and root space for different types and sizes of plants. Its lines are straight and its bottom is wider than its top in order to hold a large quantity of compost where the roots are feeding and also to facilitate easy removal of the soil and root ball when planting or repotting. There are various sizes of tube pots used in our county.

Tube pot sizes	Height (cm)	Diameter (cm)	Cost per pot (Rs.)
Tube pot	20	13	15.00
¼ size pot	18	22	15.00
½ size pot	20	27	30.00
¾ size pot	25	32	50.00
Full size pot	35	35	65.00
Tub size pots	35	50	90.00

2. Seed pan and seed boxes

A seed pan is a shallow earthen pot with a top diameter of approximately 35 cm and a height of about 10 cm. Drainage is either provided by one hole in the center or by three holes equidistant from one another. The seed boxes are made from wood, measuring 40 cm long, 60 cm wide, and 10 cm deep. The bottoms have 6-8 evenly spaced holes.

Each hole has a crock placed against the concave side. A few large pieces of crock are placed over it and also 2 or 3 handfuls of coarse sand are sprinkled at the side of this crock to prevent fine soil from clogging the drainage. Next, the necessary soil mixture is added. Seeds such as Cineraria, Begonia, *etc*. should be sown in these containers as they are quite delicate.

3. Polythene bags

For propagating cuttings like Jasmines, Durantas, Crotons, *etc*. small polythene bags with drainage holes and porous rooting medium are placed in the mist chamber. Many times, young seedlings raised in nurseries are then transplanted into polythene bags and kept there till they have attained the necessary growth to be planted in the main field (Papaya, Curry leaf, *etc.*).

4. Plastic pots

A plastic pot is a round or square container used mostly for keeping indoor plants. In addition to being reusable, light in weight, non-porous, and easily storable, they also require little storage.

Tools and implements for nursery work

Rose can/water can	:	It is used for watering the nursery. Small seeds should be watered with fine sprays of water.
Digging fork	:	It has 20 cm long prongs attached to a wooden handle. It is used to uproot plants, root cuttings and harvest tubers without damaging the roots or tubers.
Shovel	:	A curved steel plate attached to a wooden handle is used for transferring soil, manure, *etc.*
Garden rake	:	It is used to level lands and collect weeds. A rake consists of a crowbar with long handle and several projections like nails.
Hand trowel	:	The hole-making tool is used for planting seedlings and small plants. It may also be used to remove surface weeds from nursery beds.
Secateur	:	In fruit trees, this is used to control shoot growth by cutting small shoots.
Budding or Grafting knife	:	It is used for budding and grafting. In a budding operation, one of the blades has an ivory edge that lifts the bark.

Potting

Plants are potted in order to serve the following purposes:

1. Plants such as grape rooted cuttings are prepared for sale.
2. Crotons are plants that are grown for decoration.
3. Growing plants for experiments, such as pot culture studies.
4. For use as rootstocks in certain grafting methods, such as inarching mango trees.

Pot mixture or potting compost

It is necessary for potting plants. Various ingredients are used to make the pot mixture. The proportions of the pot mixtures vary based on the type of plant.

1. Ideally, the pot mixture should have an open structure that allows good drainage, holds sufficient moisture for plant growth, and allows excess water to drain away.
2. It should supply adequate nutrients to plants at all stages of growth and development.
3. It should be free from all harmful organisms and toxic minerals.
4. It should be light weight.

Potting procedure

1. Prepare the seedbed by moistening it before lifting plants. Lift the plant with a ball of earth that has as much root system intact as possible. Avoid pulling out seedlings in the hot sun. Do not allow the roots or soil around the roots to dry out.
2. Pots should be filled by first placing some crocks, then a layer of sand (5 to 8 cm) and finally pot mixture (8 to 10 cm).
3. Place the plant with the ball of earth in the center over the layer of pot mixture (In the case of root stock plants used for inarching, place on one side of the pots).
4. Press the mixture around the ball of earth, fill it up and level it off, leaving an inch head space at the top. Do not press the mixture over the ball of earth. The roots will be broken and damaged.
5. Plant stems should be at the same height as they were in the seed bed.
6. Plant pots should be submerged in water gently until bubbles stop escaping. Remove the pot and place it under shade.

Repotting

The purpose of repotting is to change the soil medium for potted plants.

Pot bound condition

Potted plants grown over more than one season or year in pot become a tangled mass of roots very quickly, exhausting all the nutrients in the pot's limited soil. Besides, it is circumscribed in the limited place. This stage is called pot-bound.

Repotting procedure

1. A potted plant should be rest 24 hours earlier in order to facilitate repotting (removing the plant from its pot).

2. The technique for removing the plant with a ball intact is to hold the bowl with the right hand palm over the soil, placing the stem of the plant between the first two fingers, turning the pot upside down, holding the pot at the bottom with the left hand, gently knocking the rim of the pot against any hard surface such as a table or even another pot. The ball of earth comes out of the pot. In case it does not come out, break the pot with a stone or fork and knock the sides to remove the soil.
3. Examine the roots and cut neatly the decayed, dead and dried or twisted roots with help of a secateur. Reduce the size of the ball of earth surrounding the roots.
4. Plant the new pot at the same height as the old one. Submerge pot in water after filling with new pot mixture.

General

1. After potting and repotting, the plants initially wilt. Transpiration loss needs to be checked to help plants recover. Freshly potted plants should be kept in shade and watered daily.
2. The plants should be gradually exposed to sunlight after about ten days under shade by keeping them for a few hours under the sun and then putting them back in shade. By increasing the exposure every week, the plants can eventually be kept in the open. This process is called "hardening".

Reference

Kumar, N. 1997. Introduction to Horticulture. Rajalakshmi Publications, 28/5 – 693, Vepamoodu Junction, Nagercoil. pp.15.47- 15.50.

11

Nursery Management Practices

M.L. Jat, R.K. Jat and Pankaj Yadav

Introduction

Management is the process of successfully accomplishing tasks with and through others. It is all about having more control and getting better results as a result of having more control. Many parts of nursery administration are unrelated to horticulture in any way. Nurseries, both wholesale and retail, are frequently managed by persons who have worked in a range of sectors. Nutrient management, disease control, irrigation management, insect pest management, and marketing management are just a few of the nursery management techniques that are necessary for a successful nursery output. A nursery manager and employees who spend a lot of time for potting plants, weeding plants, or conversing with clients may discover that they do not spend enough time for managing the nursery, resulting in a loss of control. Many activities are carried out during management, such as planning - need for planting material, provision of mother blocks, land area requirements, water supply, working equipment, growth structures, and input availability. Implementation - land treatment, biotic interference and soil erosion prevention, correct layout, input supply, and so forth. However, in a small nursery where several jobs are required, the manager must strike a delicate balance between the numerous responsibilities. Managers who are well informed can provide good management; thus, the first duty of every manager is to become knowledgeable about the company for which they are responsible.

An effective nursery management technique includes a management flow plan that makes it simple to control nursery production. To get a better understanding of the methods involved in nursery production, create an operational flow plan that illustrates the numerous actions taken in each of the four stages (propagation, transplanting, growing on and marketing). A flow plan may benefit many aspects of nursery management, including production efficiency, quality control, finances, and deadlines. Staff can also use flow plan to help with normal management and maintenance processes.

Operational flow plan for seed propagation in different stages

1. Propagation stage: Several actions are carried out during this stage:

- The seeds have been taken.
- Combine the propagation media (vermicompost, perlite, sand and vermiculite).
- Clean and sterilize propagation media (if applicable).
- Filling the seed tray with seeds.
- Seeds are sown.
- The seed is covered.
- Arrange the trays in the propagation area (*e.g.* greenhouse).
- The process of germination begins.

2. Transplantation stage: This is second stage of propagation. During this stage many activities are followed:

- Prepare the growing medium (mix soil with other media).
- Use sterilizing agents to sterilize the medium (Formalin).
- Place the seedlings and soil mix in the potting area.
- Seedling transplantation.
- For growing on, potted plants are sometimes relocated to a sheltered or semi-protected location (*e.g.,* a shaded location).

3. Development stage: This is third stage of propagation. During this stage many activities are followed:

- Buy potting soil or mix and sterilize it.
- In the potting area, mix the dirt and plants together.
- Put into a container that will be used to sell the plant.
- Allow for complete growth (apply fertilizer and water, prune *etc.* as needed).

4. Marketing stage

- Prepare and label for sale.
- Put everything in the van.
- Make sales calls to retail nurseries.
- As plants sell, unload them.

This is only one technique through which a nursery may manufacture and sell seedlings. A flow plan of operations, regardless of the technique you employ, can be useful since it highlights the procedures in each stage of production and makes them easier to analyze.

Plan for increase operational efficiency

To assess the operational efficacy of the nursery management flow plan, evolution is required. The efficiency of nursery management methods may be critically evaluated by thoroughly analyzing them step by step. Find areas where the operation is effective and areas where it can be simplified through evolution. This is beneficial in terms of saving time. Operational efficiency may be improved in a variety of ways, such as eliminating media mixing by purchasing pre mixed media, or eliminating potting up by directly sowing into the container in which the plant will be sold. When you use a flow plan to evaluate operational efficiency, you may concentrate on the areas where you can make the most progress. Additionally, a flow chart can be enhanced by adding a column that indicates how long each task takes. This may be used to figure out which tasks take the most time, how much each job costs in terms of labour, and what kind of labour is required for each work.

Plan for quality control

The use of a flow plan can also help with product quality control. The flow plan aids in the identification of all inputs and the evaluation of all techniques that may be used in the production system. The flow plan allows for more thorough quality control processes to be implemented. Some big nurseries keep meticulous records, such as the amount of nutrients given in fertilizers or the amount of light received by plants. A flow plan assists in monitoring these aspects as well as identifying the cause of any problems and resolving them. Using a flow plan in this fashion entails not just identifying the phases in the manufacturing process, but also itemizing the inputs and outputs for each step. The flow plan structure is important for this since it enables for the implementation and tracking of particular quality control processes at each stage of production.

A. Nursery water management

Every day, growers must make decisions about water management, such as when to irrigate, how much water to apply, which plants to irrigate, and how to maximize efficiency. There are several municipal and state water laws that the farmer must follow, and it is the grower's job to plan and manage their water sources to ensure compliance.

A nursery management system must include a water management system. Not just to guarantee that plants receive sufficient water, but also to keep the danger of overwatering and water waste to a minimum during water management. The water management plan that considers climate conditions, plant species, disease concerns, water application rates, and whether plants are grown in containers or in fields. When choosing a nursery location, ensure that there is always enough water and that it is of excellent quality.

1. Water quality

Both during propagation and subsequently when growing on in the nursery, water quality has an impact on the quality and rate of plant development. A nursery's water supply may be good now, but it may not be the same next month. Water quality should be checked on a regular basis, and if the quality is questionable, the intervals should be shorter. All sorts of pollutants, like salts or fertilizer residues, dust or soil particles, as well as biological contaminants such as parasites and diseases, should be kept out of nursery water.

a. **Salinity:** When there are significant quantities of salts in the water, it is called salinity. An EC (electrical conductivity) meter, available from hydroponics vendors, may be used to assess salinity.

b. **Hardness:** Hardness refers to the presence of high amounts of calcium or magnesium in water. This makes soap lathering difficult and leads to the production of scale or crust, which reduces the efficiency and longevity of pipes and water heaters.

c. **Contamination:** If water from the nursery or other properties enters the storage, it may be polluted with fertilizers, pesticides, or even diseases, which may then be distributed throughout the nursery when you water. Algae and other weeds will grow in shallow water that is exposed to light and contains some nutrients. If water quality is to be preserved, dams must create an adequate biological balance of animal and plant life.

2. Water treatment

A nursery may need to treat water if it is contaminated due to physical, chemical, or biological problems. Some nursery producers collect, recycle and utilize waste or runoff water. For many nursery operators, this is quickly becoming a legal and environmental need.

Different methods for water treatment

1. Using an active system to remove or kill plant pathogenic fungus and bacteria using chemicals, heat, or microfiltration. The microflora is also altered or destroyed by this approach. Chemical residues can impact water quality by raising salinity levels.
2. Through a passive system that does not alter the chemical characteristics of the water and maintains a steady population of microorganisms that inhibit disease organisms.

Some other methods for water treatment

Water treatment procedures include oxidation, chlorination, fluoridation, UV radiation, microfiltration, gradual sand filtration, oxidation and recycling.

i. Oxidation

In terms of water treatment, oxidation procedures are considered one of the most effective methods for removing bio-recalcitrant organic pollutants and the eradication of pathogen microorganisms that cannot be treated by conventional methods (Garrido-Cardenas *et al.*, 2019). Installing bubblers or fountains in the water improves oxygen and decreases algae development. Aerated water, according to certain propagators, is superior for propagation.

ii. Chlorination

In many crops, polluted irrigation water is one of the main sources of *Phytophthora* and *Pythium* disease inoculum (Hong and Moorman, 2005). In ponds, rivers, canals, streams, lakes, runoff water, watersheds, reservoirs, wells, holding tanks, effluents, flow, recirculating and hydroponic systems (Hong and Moorman, 2005) found 17 *Phytophthora* species, 26 *Pythium* species, 27 genera of fungi, eight species of bacteria, ten viruses, and 13 species of plant parasitic nematodes.

Unless the chlorine concentration is more than 0.4 ppm, chlorination has little effect on crops. It is a cost effective way of disinfecting water and is still used to treat municipal water (Havard, 2003). Some growers have already started using this technology to clean up their irrigation systems and water.

However, precise suggestions for usage in nursery irrigation to limit the spread of plant diseases have not been adequately examined (*i.e.*, critical free chlorine concentrations and contact durations for both plants and pathogens).

Chlorine gas, chlorine dioxide, calcium hypochlorite, and sodium hypochlorite are some of the most often utilized chlorine compounds.

iii. Fluoridation

Fluoridation does cause crop damage in rare cases. Fluoride is a common phytotoxicant that is found in large amounts in the air, water, and soil. Physiological effects of fluoride on plants have been documented, including reduced plant growth and development (Elloumi *et al.*, 2005; Jacobson *et al.*, 1966), chlorosis (Mcnulty & Newman, 1961), leaf tip burn, and necrosis (Hadujue, 1966). Some green plants, such as species of the genera *Chlorophytum, Cordyline*, and *Dracaena, Maranta, Kerchoviana,* and *Spathyphyllum*, will be harmed by the 0.5–1 ppm added by some local governments to decrease tooth decay.

iv. Ultraviolet radiation

This method only works if the water is clear enough for UV light to pass through.

v. Microfiltration

Micromesh filtration may remove suspended particles.

vi. Recycling of water

Designing and implementing an effective irrigation scheme, or improving the efficiency of an existing system, is the most essential component of saving water and decreasing water runoff and pollution from nutrient leachate. When designed properly, a water-wise system will reduce wastewater volume as well as runoff, nutrient leachate, and the need for a recycling system.

Operators of nurseries should be aware that legislation might compel them to manage, if not modify, production methods and activities that pollute waterways. This includes the need for wastewater collection, storage, and treatment. As a result, it makes sense to start with a programme that uses as little water and nutrients as possible. Nurseries that have adopted a recycling scheme have seen a 50 per cent reduction in their water consumption. With the usage of capillary mats, this figure may be lowered by another 20 % and by another 10 % if capillary mats are used in conjunction with drippers.

A nursery's apparent inconsistency between its requirement to grow excellent plants on a low cost system while also implementing an ecologically friendly production method may be substantially resolved by employing a recirculating system customized to each nursery's needs.

- Sealed, well-drained beds that collect irrigation water.
- Collection ditches at the bottom of the beds.
- An appropriate storage facility, such as a tank, pond, or dam, depending on the size of the nursery.
- An irrigation scheduling system designed to reduce water loss and capture as much recycled water as possible.

Nutrients are also necessary to investigate nutrient usage in nurseries. It is important that nurseries implement a programme that reduces the use of fertilizers, especially nitrogen, to prevent nutrients from leakage into wastewater. As a result, the wastewater collected for recycling will be of higher quality and will require less treatment before being reused. Due to dilution, recycled water contains very low levels of pathogens in nurseries where high volumes of overhead irrigation water are utilized, and recycled water is kept in big capacity dams that receive huge volumes of water on a regular basis. The biological activity is aided by reeds and other water plants along the banks, but no chemical treatment is required.

3. Different methods of watering in the Nursery

Many factors must be considered when determining the optimum way of watering in the nursery plants, including the quantity of plants being produced, the type of plants, labour costs, setup expenditures, maintenance, and water use efficiency. It's often preferable to mix and match different types of watering systems based on the demands and requirements of the specific plant species being cultivated.

I. Hand watering

Typically, hand watering is the most feasible irrigation technique for tiny native plant nurseries or nurseries growing a variety of species with different water needs. In the early stages of a nursery, hand watering is usually the best method. After understanding the various watering requirements for each nursery crop, an investment in appropriate irrigation systems to satisfy the plant's demands can be made if desired. Hand watering requires only a hose, a few types of nozzles, and a long-handled wand; there is no need for anything else. A few minor improvements, such as overhead wires to direct the hoses and rubber boots for the workers, will make the watering task more comfortable and efficient (Biernbaum, 1995). Although the work appears simple, it requires good skill and the delivery of the correct quantity of water to a range of native plant species in a variety of containers and at varied stages of growth. Irrigators should have a responsible attitude and be adequately trained to operate efficiently with water application, according to nursery management. The following are some good hand watering techniques:

- Allow water to reach the plant's roots.
- To save water and prevent foliar infections, avoid spraying the leaves.
- An angled watering nozzle should be used to avoid washing out seeds, medium, or mulch.
- For young germinant, use a thin, gentle spray; for bigger plants, use a higher volume nozzle.
- To irrigate efficiently without wasting water or compacting or washing away the root media, adjust the flow, volume, and rate of watering (Biernbaum, 1995).
- Ensure consistency in water distribution so that all plants are adequately watered; account for microclimate differences in the nursery (for example, plants on the outer edge of a south facing wall may need more water).
- Ensure that each crop is adequately watered according to its stage of growth; over time, acquire a "feel" for crop watering needs.

Advantages

- It is necessary to choose low cost and easy to install equipment.
- It is adaptable and can accommodate a variety of species and container sizes.
- Irrigators are near the crop on a regular basis and can see infections or other possible concerns.
- It allows water to be channeled beneath the leaves of plants and minimizes disease risk.

Disadvantages

- It takes a long time and requires a lot of effort.
- It is a daily duty because plants do not have weekends or vacations.
- To accomplish it right, you will need expertise, experience, and presence of mind (the task should not be assigned to inexperienced staff).

- There is a risk of the medium being washed out or compacted from the containers.
- There is a chance of runoff and trash.

II. Fixed overhead sprinklers

Fixed overhead sprinklers consist of a grid pattern made up of a series of parallel irrigation lines, generally made from plastic polyvinyl chloride tubing, and placed at equal intervals. Sprinklers that spray water from the ceiling perform a good job if they are properly built and maintained. In accordance with the number of simultaneous nozzles that the pump is able to operate at the required water pressure, the propagation environment is split into irrigation "bays" or "zones." The ideal working pressure varies as per sprinkler type, and manufacturer standards are provided. Because some sprinklers have multiple coverage options, such as full circle, half circle, and quarter circle, irrigation lines can be placed around the perimeter of the irrigation bay to provide full overlap coverage. It should be possible to control each bay independently by connecting a solenoid valve to a timer that will program the lengths and sequences of irrigation. Each irrigation bay may be customized in size to accommodate species with varying water requirements within a larger growth structure. When constructing a new irrigation system, it is a good idea to get the aid of an irrigation expert to ensure that the coverage and water pressure are matched. Fixed overhead irrigation systems employ a variety of spray nozzles. In spinner sprinklers, which have offset nozzles on rotating arms, water pressure drives them to spin in circles. Stationary nozzles have no moving parts and distribute water in a circular pattern; half circles and quarter circles are also available.

On occasion, overhead irrigation lines are also equipped with mist nozzles, which are primarily used to control humidity during the germination process. Sprinklers in the basal area (fixed basal sprinklers) large outdoor growing or holding areas frequently employ basal irrigation systems. They utilize a regular grid of permanent or moveable irrigation lines with evenly spaced sprinklers, similar to overhead irrigation systems in design and function. Sprinklers with both fixed and rotating-impact nozzles are widely utilized. These sprinklers rotate slowly due to the impact of the spring-loaded arm as it moves in and out of the nozzle stream. Many manufacturers provide rotating-impact sprinklers with various nozzle diameters and coverages. Water pressure is especially important for proper distribution of water from these sprinklers as they are powered by the pressure from the nozzle jet. Due to the high coverage area of impact sprinklers, less irrigation tubing and fewer nozzles are required with basal irrigation systems. Irrigation Systems with Moveable Booms The movable boom as, which delivers water in a straight pattern and solely to the crop, is the most efficient but most costly kind of sprinkler irrigation. Most small native plant nurseries believe moveable booms to be too expensive, although they should be explored wherever possible (Landis *et al.*, 2009)

Advantages

- Designing and installing them is relatively easy and affordable.
- There are several nozzle designs and application speeds to choose from them.
- A "cup test" can be used to determine water distribution patterns.

Disadvantages

- For large leaved crops, foliar interception renders overhead watering inefficient.
- Due to poor circular irrigation designs, irrigation water might be wasted.
- Excessive water on the leaves might raise the risk of foliar diseases.
- The nozzle drop from leftover water in lines from above sprinklers can damage germinates and young plants.
- Irrigation lines for basal sprinklers must run down the floor, posing a hazard to employees and equipment.

III. Overhead irrigation systems

Container nurseries have long preferred overhead watering systems since they are very inexpensive and simple to install. With native plants, especially those with wide leaves, the inherent inefficiency of overhead systems becomes a major concern. Wide leaf blades, along with the tight spacing of most containers, produce a canopy that intercepts the majority of the water delivered by overhead irrigation systems, lowering water usage efficiency and causing inconsistent water distribution among individual containers. Subirrigation systems, which provide a viable option for native plant nurseries, can avoid these issues. It is a relatively new irrigation method. This technique has been used for years to produce wetland plants, but it is also being used to produce forbs (Pinto *et al.*, 2008), conifers (Dumroese *et al.*, 2006) and hardwood trees (Dumroese *et al.*, 2006; Davis *et al.*, 2008). Subirrigation systems involve submerging the bottoms of containers with water on a regular basis (for example, for a few minutes once a day). After the water drains out, the growth media become completely saturated, but the leaves remain dry. Subirrigation avoids the issue of big leaves intercepting overhead water, as well as other issues associated with overhead irrigation.

Capillary action is used in all subirrigation systems to transport water up through the growth media against gravity. The attraction of water molecules to one other and other surfaces causes capillarity. Water will travel up through the growth media in the containers after the subirrigation tray is flooded, with the degree of this movement dependent on the properties of the container and the growing medium, primarily the latter. The narrower the holes between the particles of the growth media, the higher the water will rise. The water drains out once the root systems have been saturated (typically a few minutes). Subirrigation is the process of replenishing moisture in the growth medium by supplying water from the bottom regularly. Several alternative subirrigation techniques have been devised, however some, such as capillary beds and mats, will not function in native plant nurseries' narrow bottomed containers. Others, on the other hand, show potential. Containers are typically laid on the floor in a shallow structure made of pond liners surrounded by a raised border of wood or masonry, for instance, in ebb and flow or ebb and flood systems. After the growth material in the containers has been saturated, subirrigation trays, troughs, and bench liners are filled with water and drained. For a wide range of native plants, either of these subirrigation methods should suffice. Although commercially accessible prefabricated subirrigation systems are available, nurseries on a tight budget may consider building their own using readily available components. The construction of a trough system is possible using concrete blocks and plastic liners or prefabricated drainable ponds made of plastic.

IV. Capillary mats

Capillary mats are water absorbing mats used under containers. Water is poured into the mat, where it is gently absorbed by the soil in the containers. These are getting increasingly popular because they consume around 60 % less water than sprinklers. The disadvantage is that without frequent cleaning of the potting media with clean water, salt and mineral deposits can build up in the container and burn the plant roots.

V. Micro irrigation

Micro irrigation may be a particularly effective way of water delivery for nurseries that grow plants in 1 gallon (4.55 litres) or bigger pots. Micro irrigation often uses poly pipe with micro sprayers (also known as "spitters" or "spray stakes") or drippers put into each container separately, with tiny tubing used to extend the emitters outside the poly tube. Micro sprayers are frequently chosen over drippers because they moisten a larger surface area and uniformly distribute water throughout the container. A sprayer's operation is also easier to visually evaluate than a dripper's. Filtration is required for micro irrigation systems to prevent emitters from becoming clogged. Because micro irrigation systems have a sluggish infiltration rate, each irrigation station will have to operate for a long period in order to provide enough water to the plants. Likewise, if a container dries out, hand watering to rewet the growth media may be necessary before drip irrigation can work.

Advantages

- It is important to give water directly to root zones (not to foliage, which can cause disease).
- Water is used very efficiently; less than 10% of the water applied is wasted.
- The delivery is consistent; each container receives an equal quantity of water.
- The rate of infiltration is satisfactory (due to a slow delivery process).
- Leachate production is also decreased.

Disadvantages

- It is difficult and time consuming to design a system and install each emitter separately for each plant.
- Installing for plants growing in pots less than 1 gallon is typically inefficient.
- Because of the sluggish water delivery, each irrigation station must run for a long period.
- The emitters are easily plugged (water filtration and irrigation system maintenance are necessary).
- It is difficult to physically check for water supply, and issues are frequently overlooked until it's too late.

4. Water conservation and managing nursery wastewater

Nursery runoff and wastewater may be key variables to address depending on the irrigation system's effectiveness. Sprinkler irrigation from above is ineffective. Irrigation systems that use micro irrigation or subirrigation are highly efficient, although they are inconvenient for some containers and plants. Considering that many container nurseries apply insecticides and fertilizer through the irrigation system, low irrigation efficiency is more than just a waste of water. To encourage the leaching of surplus salts, liquid fertilizer is generally given in excess of the amount required to saturate the growth medium. The majority of pesticides are administered by irrigation in a water based carrier, and some of these chemicals eventually wind up in wastewater runoff; growth media drenches are particularly problematic in this regard. Originally, it was assumed that fertilizer salts and pesticides were filtered out or absorbed by the soil, but this theory has since been debunked. Several studies have found that excess fertilizer nutrients and pesticides leach from containers at conifer nurseries, contaminating groundwater.

The most efficient method to deal with the problem is to increase the irrigation system's efficiency and employ water saving techniques, such as mulching. In addition to collecting runoff for treatment, it can be diverted to other plants for absorption as it occurs. Irrigation design and application that is effective and efficient saves water while fulfilling the demands of the plants. In the nursery, there are several horticultural methods that can assist to conserve water and minimize water consumption and runoff. Due to the cost of water treatment and the risk of reintroducing extra salts or pests, direct recycling of spent nursery water is rarely done on a modest scale. However, in regions where water is scarce, these types of capture and recycle systems for nursery runoff water may be economically viable. A less high tech method for water reuse is to collect water runoff from the nursery using an impermeable nursery floor (*e.g.*, pond liner). The water can be stored in tanks or used directly to water other crops. The salt tolerant plants, may benefit from runoff water, and these plants will also clean and filter the water. Nurseries that cultivate aquatic or semiaquatic plants may be able to divert runoff to these plants, increasing the nursery's water use efficiency. Nursery runoff water can be used to irrigate crops in the field, such as seed orchards, wetland crops, adjacent landscaping, or tree crops. In most situations, an extra sand bed or filter is utilized to strain the particles out of the water before it is sent to the crops *via* an irrigation system. The system can be gravity fed if the crops are placed downhill from the nursery. A pump will be required to administer the water to the crops otherwise.

B. Plant protection in nursery

A plant is sick when it is unable to operate normally. Pathogens and environmental factors are the major causes of disease in trees. Viruses, bacteria, fungus, nematodes, mycoplasma-like organisms, and parasitic higher plants are all disease pathogens found in trees. The most common pathogens are fungi. Seed rot, seedling damping off, root rot, leaf diseases, canker, vascular wilt, dieback, galls and tumors, trunk rot, and aged tree degradation are all caused by them. Unfavourable weather and environmental conditions such as extremes in temperature and moisture, severe winds, or ice can cause direct harm to trees and predispose them to insect assault.

The following is a quick summary of the different diseases and associated nursery management measures:

1. Damping off
2. Leaf web blight
3. *Colletotrichum* leaf spot and blight
4. *Alternaria* leaf spot and blight
5. *Pseudocercospora* leaf spot: It is caused by *Pseudocercospora subsessilis*
6. Powdery mildew
7. Other foliar diseases

A planned approach to plant protection is necessary for proper growth of nursery plants. Regular inspection is necessary for improved protection against pests and diseases.

Identifying the problem

Generally, plants that are not growing well are caused by one of the following:

- Insects, arachnids, birds, mammals, and other creatures are examples of animal pests.
- Fungi, bacteria, and viruses are examples of diseases.
- A problem related to the environment, such as too much or too little light, temperature, humidity, *etc*.
- Nutrient deficiency occurs when a plant does not receive the proper nutrients in the right amount or combination.

Pest identification

If you suspect or know an insect is causing a problem but aren't sure which bug it is, look at how it eats. The list below may assist you in rapidly identifying insects that are causing harm to a plant. Depending on the kind of plant damage they cause, they are categorized. This is not a complete list, but it should cover most of the types:

- Insects (and other pests) which chew above ground – ants, army worm, bugs, beetles, caterpillars, crickets, cut worm, earwig, flea beetle, grass hopper, leaf miner, leaf roller, leaf skeletoniser, mice, possums, rats, sawfly, slug, snail, springtail and weevils.

- Insects (and other pests) which suck plant parts above ground – aphids, harlequin bug, lace bug, leaf hopper, mealy bug, mite, psyllid, scale, squash bug, thrips, tree hopper and whitefly.
- Insects (and other pests) which feed below ground – root aphis, root nematodes, root borer, root worm, root weevil, woolly aphids, wire worm and beetle larvae.
- Borers (including fruit borers) – codling moth, bark beetle, corn earworm, white pine weevil, melon worm, longicorn beetle, European apple sawfly.

Identifying diseases

Fungi, bacteria, and viruses are examples of plant diseases. Diseases have a wide range of symptoms, some of which are comparable to and occur in combination with insect infestations. The following are the most prevalent diseases:

- Anthracnose – dead, sunken spots
- Blight – plants part die quickly
- Canker – an accumulation of dead tissue in one place
- Damping off – rotting of young plants at soil level
- Galls – an abnormal swelling
- Leaf spot – dead or off-colored spots on leaves
- Mildew – the young growth becomes distorted and powdery-gray in color
- Rots – decaying tissue
- Rust – brown/orange spots or stripes
- Smut – sooty/powdery covering
- Sooty mould – similar to smut (associated with insects)
- Wilt – drooping foliage

Timing of disease control measures

When using chemical disease control is economically viable, such as in the case of forest nursery stock, it is critical that the pest management understands the disease's life cycle. For many diseases, there may be just a limited window of control in a calendar year, or the control spray may need to be administered before any signs or symptoms of disease appear. Chemical controls should be used on plants when infection is most likely to occur, otherwise time, effort, and money will be wasted. You will be able to make correct and timely management decisions if you understand the disease organism's life cycle.

Disease surveys

Disease surveys are essential because they are the initial step in putting control measures in place. Early disease detection, appraisal, and control surveys are conducted to get information on the scope of the assault, the amount of the damage, control options, cost estimates, and the delimitation of control zones, as well as to assess the efficacy of the control programme.

Protection of diseases and pest

Seedbed management

Damping off is the most common disease in nursery plants. Good sanitary conditions are required for its control. Treatments with 50 per cent ethyl alcohol, 2 per cent calcium hypochlorite and 0.01 per cent mercury chloride are used as preventative measures. These treatments last anywhere between 5 to 30 minutes. There are following seed treatments.

1. Disinfestants

These are the chemicals which eliminate organisms present on the surface of the seeds. Calcium hypochlorite is used to eliminate the infection on surface of the seeds.

2. Disinfectants

Formaldehyde, hot water and aerated steam are used to kill the infection within the seed. For water treatment, dry seeds are submerged in hot water between 48° to 57° C for 15 to 30 minutes. After treatment, the seeds are cooled and spread out in a thin layer to dry.

3. Protectants

These are chemicals applied to the seeds to protect them from soil living pathogenic fungi. Many chemicals are used as protectants and mainly two methods are in vogue for protection of seeds from the diseases. These are dry seed treatment and wet seed treatment methods. Dry seed treatment method gives better protection against soil borne diseases, whereas, wet method gives protection both from soil and seed borne pathogens.

In dry seed treatment method, the seeds are thoroughly shaken with chemicals in a rotating seed duster for some time so that the chemical sticks well to the seed surface. The most commonly used chemicals for this treatment are agrosan-G, ceresan. brassicol, captan, thiram and dithane-M 45. In wet treatment method, the seeds are kept immersed for a certain period in the liquid suspension of the chemical. The seeds should be stirred continuously during immersion for proper contact of the seeds with chemical and to avoid excessive accumulation of chemical at one portion of seed. After treatment, the seeds are dried in sun or shade quickly. The most commonly used chemicals in wet treatment method are agallal and thiram. However, certain antibiotics can also be used for wet treatment method.

4. Treatment of the soil

Soil includes toxic fungus, bacteria, nematodes, and even weed seeds, all of which impair plant growth and development. Heat and chemical treatment can be used to get rid of them. For this, the soil is sterilized by heating it to around 60°C for 30 minutes.

5. Chemical treatment

Among the substances, formaldehyde, methyl bromide, chloropicrin, and vapam are employed. Other diseases that have been observed include rust, powdery mildew, leaf spot, bacterial blight, and yellow vein mosaic. Bordeaux combination, carbendazime, and redomil can be used to treat various diseases. A biofungicide called *Tricoderma viride* can also be used.

Integrated nursery disease management

- Choosing seeds or propagules that appear to be healthy for seedling production.
- Dressing of seeds with 0.2 % carbendazim, methyl thiophanate, benomyl and thiram.
- The seeds should be sown in sterilized/fumigated, clean beds with adequate watering.
- During budding and grafting, use sterilized budding knives, secateurs, and scissors.
- Transplanting of seedling is done after root dip for 3-5 minutes in 0.02 % carbendazim solution.
- Maintaining healthy planting materials through proper sunlight, watering, and a clean environment.

- Regularly examine the health of seedlings and remove diseased stocks.
- Foliar spray of 0.2 % carbendazim/dithane M-45 at regular interval.

Important disease in nurseries and their control

1. Damping off

Damping off is the most common and devastating disease in nurseries, resulting in significant seedling losses. There are two types of damping off disease, known as pre emergence damping off and post emergence damping off, which are distinguished by the stage of seedling growth at which they are targeted. *Pythium, Phytophthora, Fusarium,* and *Rhizoctonia* are among the soil fungus that cause this disease, with the last two fungi being particularly common in Indian forest nurseries. High soil temperature, high soil moisture, high soil pH (alkaline), high nitrogen content, low light intensity due to shade, stiff or clayey soil with poor drainage, and intensive sowing are all factors that favour disease growth. *Fusarium oxysporum* was the cause of the disease in neem seedlings (Mehrotra and Pandey, 1992).

Control measures: The disease has been controlled through cultural techniques that promote plant development while inhibiting the growth of plant pathogens. Chemicals such as formalin and appropriate fungicides are used. Fungicides can be administered as a soil drench or a soil mix, and formalin is used as a soil fumigant. In other situations, seed treatment with thiram @ 3 g/kg of seed was also shown to be highly beneficial. Before used, sun heating should be used to disinfect the potting media.

2. Leaf web blight

This disease is caused by *Rhizoctonia solani.* After the regular monsoon rains begin, the disease appears in the nursery (Mehrotra, 1990; Sankaran *et al.,* 1986).

Symptoms: Greyish brown spots appear, which grow in size as the fungal hyphae advance, eventually engulfing the entire leaf blade. As though trapped in a spider's web, the diseased adjacent leaves are linked together by the fungal hyphae, giving rise to the term web blight. Premature detachment of the leaflets or the whole pinnae occurs. The disease is propagated by overlapping leaves coming into touch with each other.

Control measures: The disease has been handled using an integrated strategy that incorporates sanitation and cultural norms. Fungicide (bavistin @ 0.1 per cent) application is shown to be effective.

3. *Colletotrichum* leaf spot and blight

This infection is caused by *Colletotrichum gloeosporioides*. It has been found in a serious form in the New Forest, Dehra Dun. The first symptoms appear in nurseries at the end of September or the beginning of October (Mehrotra and Pandey, 1992).

Symptoms: The fungus creates leaf spots that quickly grow in size and cover vast portions of the leaf. The diseased leaves develop a blighted look and finally fall off the plant. Premature defoliation is seen in severely infected seedlings.

Control measures: Use of blitox fungicide @ 0.2 per cent twice weekly has been shown to suppress the disease.

4. *Pseudocercospora* leaf spot

The disease is caused by *Pseudocercospora subsessilis*. The disease is seen all over the world where neem grows naturally.

Symptoms: The infection spots are dark in color, with white patches intermingled. The fungus sporulates on the leaf's underside and produces conidia that are grey in color. The highly diseased leaves become pale and prematurely shed.

Control measures: Mancozeb in conjunction with brestan has been proven to be beneficial in the treatment of the disease.

5. Powdery mildew

Several genera cause the disease, symptoms of which are similar on many hosts. The most common powdery mildew fungi infecting nursery plants are, in the sexual stage, *Erysiphe*, *Leveillula*, *Podosphaera*, *Sphaerotheca*, and *Uncincula* and in the asexual stage, members of the genus *Oidium*. Most species of these pathogens are host specific often infecting only a few related plant species.

Symptoms: White powdery mycelia growth appears on the leaf surface and shoots. They covered the entire leaf lamina, giving it a grayish white appearance. Severely diseased leaves and leaflets defoliated early.

Control measures: Foliar spray of sulphur based fungicide has been proven to be beneficial in reducing disease severity.

6. Other foliar diseases

- **Bacterial leaf spot:** It is caused by a bacterium *Xanthomonas* species and *Pseudomonas* species.

- **Leaf spot and blight**: Leaf spot is caused by *Colletotrichum* species. Leaf blight and stem rot are caused by *Sclerotium rolfsii.*
- **Seedling wilt:** It is caused by *Fusarium* species. Twig canker and shot hole in leaves are caused by *Phoma* sp.

C. Nutrient management

Fertilizer application practices are focused on water quality and nutrient runoff, as well as promoting nursery stock development. Nutrient management should strike a balance between maximizing plant growth while preventing over nutrition. Excess nutrients can pollute the ecosystem if they are lost in the environment.

Diagnosis of soil before application of manure and fertilizers

1. Soil analysis

An accurate soil test can be helpful for many reasons, including optimizing crop production, preventing the contamination caused by runoff and fertilizer leaching, diagnosing plant diseases, improving the soil's nutritional status, reducing fertilizer consumption, saving money and detecting soil contaminated by lead or other toxins. The first stage in developing a sound lime and fertilization programme is to determine the pH and fertility level of the soil through a soil test. A soil test allows you to keep track of your soil and avoid shortages, excesses, and imbalances. When the soil is very moist or has just been limed or fertilized, sampling should be avoided. Distinct soils should be collected and analyzed individually because they appear to be different or have been treated differently. Areas with low growth should be examined independently as well. A soil test numerical findings are based on analytical techniques employed by individual laboratories. As a result, data from various laboratories' soil tests should not be compared.

2. Production in the field (field nursery analysis)

Soil samples from field nurseries can be taken at any time of year, but the best period to predict fertilizer needs for the next year is from midsummer to fall. When sampling, soils should be dry enough to till, and fields are generally dry and accessible in the fall. During late summer and early fall, soil pH and nutrient levels will be at or near their lowest values. As a result, fall samples are more indicative of real fertility conditions throughout the growth season than those taken in late winter or early spring. Fall sample also gives the laboratory enough time to deliver results and recommendations, allowing any necessary limestone and fertilizer to be administered before planting. The temperature and moisture content of the soils affect soil nutrient levels throughout the year. As a result, it is critical that samples be obtained at or around the same time each year so that results may be compared year to year.

3. Container production

A soilless mixture sample is examined differently than a field soil sample. The 1:2 dilution method, saturated media extract (SME), and leachate Pour Thru are three frequently used procedures for evaluating soilless medium that employ water as an extracting solution. A number indicating the amount of soluble salts from a soil test utilizing the 1:2 dilution method will be different from SME or leachate Pour Thru findings. For the 1:2 technique, 2.6 would be "extreme" (very high), "normal" for SME, and "low" for leachate Pour Thru. Likewise, nutrient levels are likely to vary depending on the method of testing. Use the interpretive data for the specific soil testing technique if possible to avoid incorrect interpretations of the results.

The deficiencies and remedies of important nutrients for better development of nursery production

Nutrients	**Deficiency**	**Correction measures for deficient nutrients**
Nitrogen	Leaves are light green to light yellow (chlorosis) with short and slender stalks.	Use of any nitrogenous fertilizer. N can easily be applied through foliar application of 1 % urea (10 g urea/ litre water) during active growth of the plant.
Phosphorus	Leaves are dark green, often developing red and purple in colour. The lower leaves may be yellow, which may turn greenish, brown and lastly black.	Use recommended doses of P. Application of DAP (2 %) or single super phosphate (1 %) is equally useful.
Potassium	Development of dead spots on lower leaves, usually at the tips and between the veins, internodes are short and leaves slender stalks.	Use recommended doses of K or foliar application of sulphate of potash (1 %). Foliar application of muriate of potash is equally useful but it should be avoided in alkaline/saline soils.
Magnesium	The lower leaves may be mottled or chlorotic, without dead spots with margins curved upward. The stalks are short and slender.	Soil application of dolomite or gypsum corrects Mg deficiency effectively. Foliar application of magnesium sulphate (0.5 %) is helpful.
Zinc	Development of dead spots on lower leaves, involving larger areas. The leaves are thick and stalks with shorter internodes, showing rosette appearance.	Soil application of zinc sulphate @ 20-25 kg/ha or its foliar application (0.5 %) helps to correct Zn deficiency effectively.

Calcium	Initially, terminal buds of the young leaves become typically hooked, which dies soon at tips and margins.	Apply lime to the soil or growing media, depending on the pH of the soil or media. Foliar application of calcium nitrate (1 %) is equally useful.
Boron	The terminal buds of young leaves are light green at base. Later, the leaves become twisted and die at the tips.	Borax (15-20 kg/ha) applied to soil or growing medium is quite effective to check B deficiency in the nursery plants. Foliar application of borax (0.2-0.5 %) is equally advisable.
Copper	The young leaves may be permanent wilted or show chlorosis. The twig of stalk just below the tip is unable to stand erect in later stages. Cracking of bark may also take place.	Soil application of copper sulphate (10-12 kg/ha) or its foliar application (0.5 %) is recommended dose to correct B deficiency in nursery plants.
Managanese	Spots of dead tissue are found scattered over the young leaves.The smaller veins tend to remain green.	Soil application of manganese sulphate (20-25 kg/ha) or its foliar application (0.2-0.4 %) is useful for correcting Mn deficiency in the nursery plants.
Sulphur	Chlorosis of young leaves with veins and tissues between veins remaining light green in colour. The stem may be thin and erect.	Foliar application of 0.5 % gypsum (5 g/litre)
Iron	Chlorosis in young leaves with major veins remaining green. The stalks of leaves are short and slender.	Application of ferrous sulphate (40-50 kg/ha) to the soil or growing medium or its foliar application (0.5 %) is recommended to correct Fe deficiency.

Manures and fertilizers application in nursery

Types of manures

As manures, animal excreta including dung, urine, straw, and other organic materials are used in crop cultivation. FYM is the name given to degraded manure. The average nitrogen (N), phosphorus (P_2O_5), and potassium (K) content of well decomposed farmyard manure is 0.5 per cent nitrogen (N), 0.3 per cent phosphorous (P_2O_5), and 0.5 per cent potassium (K_2O). Balanced feeding of nursery plants is required for the optimal performance of fruits and vegetables. Nutrients in both organic and inorganic forms can be used to create balanced nutrition.

- Organic manures
- Biofertilizers
- Inorganic fertilizers or chemical fertilizers

Organic manures

Manures are prepared by using plants and animals debris. It can be categorized as follows:

Plant and animal waste are used to make manure. It may be divided into the following categories:

1. Plant-based manures *e.g.* green manures
2. Animal-based manures *e.g.* poultry manure
3. Manures from plants and animal origin *e.g.* compost, FYM
4. Organic fertilizers *e.g.* bone meal, fish meal, blood meal

Biofertilizers

1. Nitrogen based biofertilizers- *Azotobacter, Rhizobium, Azospirillum*
2. Phosphate based biofertilizers – Phosphate Solubilizing Bacteria (PSB)
3. Microbial decomposers- *Trichoderma viridae*

Important points regarding the nutrition management in nursery plants

- The mother plants should be provided selective and balanced nourishment through soil or irrigation. The root growth will be slowed if there is too much nitrogen in the soil.
- Different types of rooting medium are utilized in nurseries. It is devoid of nutrient, therefore we must supply nourishment in accordance with the needs of the plants. Nutrition should be given special attention in the nursery during sprouting, root initiation, and plant hardening. Foliar sprays can help to restore nutrient deficit in plants.
- Organic manures, inorganic fertilizers, and biofertilizers should all be utilized together for balanced nutrition.

Fertilizer requirement of fruit nursery plants

- Red soil and decomposed FYM or compost are properly combined in the media for filling polythene bags.
- Soluble fertilizers are drenched 5-6 times, depending on development phases, at a rate of 2 g/litre of water.
- Foliar fertilizer application is done 5-6 times by spraying, depending on plant growth stage, @ 2 g/litre of water.

Manure and fertilizer application techniques

1. **Broadcasting:** Organic manure, such as FYM and compost, is spread over the beds and well mixed with a shovel or rake. After the beds have been thoroughly prepared, the seeds are planted.
2. **Ring around stem:** To apply manures and fertilizers to mature trees and plants, a ring has to be formed around the tree or plant's trunk.
3. **Fertilizer application in polybags/urea brackets:** A weeding hoe is used to apply fertilizer directly in the polybags near the stem of the plants.
4. **Fertigation:** An adequate amount of fertilizers can be combined in irrigation water and applied to nursery plants by drip, sprinkler, or soaking near the stem irrigation.
5. **Foliar Fertilization:** Spraying chemical fertilizers on the leaves to nourish the plants or feed them nutrients. Foliar feeding is also known as spray fertilization.

References

Bhujbal, B. 2011. Resource Book on Horticulture Nursery Management. The Registrar, Yashwantrao Chavan Maharashtra Open University, Nashik, pp.42 -52.

Biernbaum, J. 1995. How to hand water. *Greenhouse Grower*, 13(14): 39- 44.

Centre for Agriculture, Food, and the Environment UMass Extension Landscape, Nursery and Urban Forestry Program, Stockbridge Hall, 80 Campus Center Way, University of Massachusetts Amherst, Amherst, Massachusetts.

Davis, A. S., Jacobs, D. F., Overton, R. P. and Dumroese, R. K. 2008. Influence of irrigation method and container type on northern red oak seedling growth and media electrical conductivity. *Native Plants Journal*, 9(1): 4-12.

Dumroese, R. K., Pinto, J. R., Jacobs, D. F., Davis, A. S. and Horiuchi, B. 2006. Subirrigation reduces water use, nitrogen loss, and moss growth in a container nursery. *Native Plants Journal,* 7: 253–261.

Elloumi, N., Abdallah, F. B., Mezghani, I., Rhouma, A. and Boukhris, M. 2005. Effect of fluoride on almond seedlings in culture solution. *Fluoride,* 38: 193-198.

Garrido-Cardenas, J. A., Esteban-García, B., Agüera, A., Sánchez-Pérez, J. A. and Manzano-Agugliaro, F. 2019.Wastewater treatment by advanced oxidation process and their worldwide research trends. *Int. J. Environ. Res. Public Health*, 17(1): 170. https://doi.org/10.3390/ijerph17010170

Hadujue, J. 1966. Reaction of some relativity resistant plants to sudden increase in the concentration of fluoride exhalation. *Biolozia,* 21: 421–427.

Havard, P. 2003. Farm irrigation water safety initiative final report April 2003. Nova Scotia Agricultural College. *Hort. Nova Scotia.*

Hong, C.X. and Moorman, G.W. 2005. Plant pathogens in irrigation water: Challenges and opportunities. *Crit. Rev. Plant Sci.*, 24: 189–208.

https://vikaspedia.in/agriculture/crop-production/tips-forfarmers/nurserymanagement#:~:text=Planning%20%2D%20demand%20for%20planting%20material,layout%2C%20input%20supply%2C%20 *etc.*

Jacobson, J. S., Weinstein, L. H., McCune, D. C. and Hitchcock, A. E. 1966. The accumulation of fluorine by plants. *J. Air Pollut. Control Assoc.*, 16(8): 412-417.

Landis, T. D. and Wilkinson, K. M. 2009. 10: Water Quality and Irrigation. In: Dumroese, R. K., Luna, T. and Landis, T. D., editors. Nursery Manual for Native Plants: A Guide for Tribal Nurseries - Volume 1: Nursery Management. Agriculture Handbook 730. Washington, D.C.: U.S. Department of Agriculture, Forest Service. pp. 177-199.

Mason, J. 2004. Nursery Management, Second edition, National Library of Australia Cataloguing-in-Publication entry, Landlink press.

McNulty, I. B. and Newman, D. W. 1961. Mechanism (s) of fluoride induced chlorosis. *Plant Physiol.*, 36(4): 385.doi: 10.1104/pp.36.4.385.

Mehrotra, M. D. 1990. *Rhizoctonia solani*, a potentially dangerous pathogen of Khasi pine and hardwoods in forest nurseries in India. *European J. Forest Pathol.*, 20(6-7): 329-338.

Mehrotra, M. D. and Pandey, P. C. 1992. Some Important Nursery Diseases of *Azadirachta indica* and Their Control. In: Monograph of Neem (*Azadirachta indica* A. Juss) (Tewari, D.N., Author). International Book Distributors, Dehra Dun, India.

Pinto, J. R., Chandler, R. and Dumroese, R. K. 2008. Growth, nitrogen use efficiency, and leachate comparison of subirrigated and overhead irrigated pale purple coneflower seedlings. *Hort Sci.,* 42: 897–901.

Ratha Krishnan, P., Kalia, Rajwant K., Tewari, J.C. and Roy, M.M. 2014. Plant Nursery Management: Principles and Practices. Central Arid Zone Research Institute, Jodhpur, pp. 40-45.

Sankaran, K. V., Balasundaran, M. and Sharma, J. K. 1986. Seedling diseases of *Azadirachta indica* in Kerala, India. *European J. Forest Pathol.*, 16(5-6): 324-328.

12

Nursery Registration Act and Accreditation

Rajesh Mor and Sonu Kumar

Introduction

Nursery is an art of raising and selling seedlings, saplings, and other planting materials for use in gardens and orchards. The availability of true to type, healthy and high quality planting materials is a must for a successful and profitable fruit production. "Scion block" refers to trees planted to provide scion wood for propagating nursery stock, whereas "seed block" refers to the planting of registered seed trees that will provide seeding rootstock for propagating certified nursery stock. It consists of planting self-rooted mother trees/mother stools with the objective of providing vegetatively propagated (clonal) rootstock for the multiplication of certified nursery stock.

Hence, a suitable nursery act is needed to improve horticulture production through the use of high quality planting materials. A nursery act has yet to be passed in some states. Several states have nursery legislation, but it has not been implemented effectively, and it should be evaluated and revised to make the necessary changes. As with other major industries, the nursery industry must comply with certain laws and regulations. It is primarily for the protection of people that laws and regulations are promulgated by government.

Model Nursery Regulation Act

In 1954, the Indian government enacted the Model Nursery Regulation Act. In Model Nursery Regulation Act some rule are made to provide good, healthy, true to type, disease free, virus free and injury free planting material to farmers/growers. Based on this comparison, some states have enacted their own nursery registration statutes to ensure that farmers have access to authentic planting material based on the locally available crops.

The act may later be known as the Fruit, and Ornamental Plant Nurseries (Regulation) Act, 1995. The Act shall come into force on the date specified by the Government in the Official Gazette.

Some definition

Appellate authority: One who is appointed in the Gazette by notification if he or she falls below the rank of Secretary to the Government for the purposes of Section 12.

Competent authority: A person or authority appointed in accordance with section 3 of the Act. In the Official Gazette, the Government may appoint the Competent Authorities. They will be appointed as follows:

- To appoint Gazetted Officers of the Government as competent authorities for the purposes of this Act; and
- Describe the limits within which the competent authority may exercise the powers and perform the duties conferred upon it by or under this Act.

Director: Means the Director of Agriculture of the Government.

Inspecting officer: An officer not below the rank of Agricultural Officer or Zonal Agriculture Officer authorized by the Director to inspect nurseries.

Notification: A notification published in the Official Gazette.

Nursery: Refers to any nursery or tissue culture unit where fruit and/or ornamental plants are grown for business purpose, propagated, and sold for transplantation or cultivation, but does not include government owned nurseries.

Nursery stock: It is defined as all trees, shrubs, brambles, woody vines, woody florist stock, herbaceous perennials, vegetable plants, bedding and other annual herbaceous plants, their roots, cuttings, grafts, scions, buds, fruit pits, seeds and their parts for propagation, except bulbs, field crop seeds, vegetable seeds, flower seeds, and other annual herbaceous plants, cuttings, grafts, scions, buds, fruit pits.

Nurseryman: A person who produces and sells fruit and ornamental plants.

Owner: The person who, or the authority that, controls the affairs of a fruit and ornamental plant nursery, then in those cases where the management, workings and/or decisions of such nursery are in the hands of the manager, managing director or managing agent, then such manager, managing director or managing agent shall be the owner of the fruit and ornamental plant nursery.

Prescribed: By means of rules enacted under the Fruit and Ornamental Plant Nurseries (Regulation) Act, 1995.

Why

- Horticultural, plantation, and other crops with low productivity due to low quality planting material.
- Uncontrolled transport of planting materials.
- Lack of checking of quality of seeds.
- Un-availability of experienced persons and skilled labour for nursery work.
- For production of virus free, diseases free and injury free planting material.
- Lack of maintains of record of nursery activities by owners of the nursery.

Procedure for grant of license

1. To obtain a license, one must submit an application form and fee to the Department of Agriculture, along with the required documents.
2. After receiving such an application, the responsible authority will do whatever investigation it deems appropriate, and if he is satisfied, he will grant the request.
    a) That the nursery for fruit and decorative plants must be suitable for ensuring the proper multiplication of the fruit and ornamental plants for which a license has been filed.
    b) That the applicant is qualified to operate or develop a fruit and ornamental plant nursery of any kind.
    c) That all other requirements imposed on the applicant must also be met.
    d) That the license fee and security have been paid by the applicant if any, shall issue the license in the prescribed form.
3. After receiving a nursery industry license application, the department or a representative will inspect all plants, plant material, or nursery stock located or growing at the business location or any other suitable location.
4. Inspections may be conducted yearly or as needed and with or without notice by the department.
5. After a successful inspection, the department will issue a nursery stock certificate to the licensed nursery.

6. If the competent authority is not satisfied, it may refuse to issue the license after giving the applicant a reasonable opportunity to be heard, after noting the reasons for such denial, and shall provide a copy of the order made there under to the applicant.
7. Every license issued under this section is valid for five years (state wise) from the date of issue and may be renewed by the competent authority from time to time upon application and payment of the applicable fee in accordance with the conditions prescribed.
8. For example Act No. 15 of 1973, established the Himachal Pradesh fruit nurseries registration rules. To apply for a license to establish a fruit nursery, the applicant shall submit Form-1 along with a Treasury Challan in the original of Rs. 100/- in favour of the concerned state Director of Horticulture, according to the act, in any Government Treasury/State Bank of India. After completion of all codal formalities, the license is initially issued for three years and can be extended further by submitting a renewal application to the competent authority 90 days before to the expiration of the previously issued license.

Responsibility of licensee

- Allow to multiply only those varieties of fruit and ornamental plants listed in the scion or rootstock license for propagation and sale by the competent authorities.
- In a method permitted by the Government, the nurseryman shall keep track of the time of various operations/applications, as well as nursery stock of various crop varieties.
- The department must approve the planting location, which must be in an area with minimal risks of infection or pest spread due to drainage, flooding, irrigation, or other causes.
- Nursery stock for fruit trees must be developed on rootstocks from certified seed trees or from registered stool beds.
- Plant certified nursery stock on soils that have been treated with a pre-plant nematicide or that have been checked for viruses, vectors, and nematodes before planting.
- When nursery stock meets the requirements, it must be labelled with the variety, interstock, and rootstock designations, as appropriate.
- Every nurseryman should keep a layout plan showing the location of every rootstock and scion.

- A nurseryman should display on a label every variety of fruit and ornamental plant that is ready for sale in an obvious manner.
- A nurseryman should maintain a register containing the name of the fruit and ornamental plant sold, its age, the species of rootstock and budwood, as well as the name and address of the person who purchased it, and should provide the record for inspection upon request by the Director or Inspecting Officer.
- The nurseryman must quarantine infected trees because there is a risk of the insect, pest, or disease spreading to other nurseries and nursery plants and do not use their budwood for future propagation.

Refusal/ cancellation of license

A license granted or renewed under section 5 may be suspended or revoked on the following grounds:

- If the nursery is not operating in accordance with the norms and regulations of the state government or the Government of India.
- When the maximum rate or price for a notified fruit or decorative plant has been determined by the Government. In the past, he has sold such fruit or ornamental plants at a higher price after receiving government notification.
- When the plants are determined to be virus infected or if the nursery has off-type plants.
- After periodic indexing, a virus infected mother tree is found.
- A pest control inspection may not be carried out properly if regulations are not followed.
- When the plants haven't been cared for properly.
- When registration numbers were misused.
- Nurseryman did not comply with any of the terms and conditions.

Important actions

Return of license: After the license expires or upon receipt of an order suspending or canceling the license, the licensee must return the license to the competent authority.

Issue of duplicate license: If a license granted to an owner is lost, destroyed, mutilated, or damaged, the competent authority will issue a duplicate license upon application and payment of a prescribed fee.

Appeals

1. A person aggrieved by an order of a competent authority refusing to grant, renew, or suspending or canceling a license, may appeal in such form and manner, within such period, and to the appellate authority as notified by the Government: provided that, the appellate authority may entertain the appeal after expiry of the prescribed period, if it is satisfied that the appellant was prevented by sufficient cause from filing the appeal in time (Fruit and Ornamental Plant Nurseries (Regulation) Act.
2. As soon as the appellate authority receives an appeal under sub-section (1), it will, after giving the appellant a reasonable opportunity of being heard, pass judgment on the appeal.
3. In accordance with section 13, the order issued under this section shall be final.

Civil penalties

1. Any violation of any provision of this Act or any rule enacted in compliance therewith, which is punishable under this section; or any violation of the Fruit and Ornamental Plant Nurseries (Regulation) Act.
2. Restriction/Obstruction to any officer or person in the exercise of any powers conferred or in the effectiveness of any duty imposed upon him by or under this Act, he shall be punished with a fine of up to one thousand rupees, or imprisonment for a term of up to one month, or both.

Power to make rules

1. The Government may make rules for implementing the purposes of this Act by publication in the Official Gazette, subject to the requirement of prior publication.
2. It may provide for all or any of the following matters, without prejudice to the generality of the foregoing power:
   a. How to make an application for a license.
   b. Under section 5, the fee and security deposit for granting licenses and renewing them, the period for which a license may be granted, the conditions under which it may be granted, and the form in which it may be granted.
   c. If the licensee violates any of these conditions, the license will be suspended or revoked under section 9.

d. A method for determining the age of fruit and ornamental plants kept for sale.

e. A register must be maintained.

f. By naming the insects, pests, and diseases that progeny trees should be protected from.

g. A duplicate license fee is payable under section 11.

h. A description of the form and manner in which an appeal can be made under section 12, as well as the procedure the appellant authority must follow in disposing of the appeal.

i. Any other matter that is to be prescribed.

Merits

- Availability of quality planting materials.
- Availability of diseases, virus and injury free planting materials.
- Increases income of farmers.
- Availability of registered and certified seeds of different variety for raising quality rootstocks.
- Increases productivity of orchard and reduce juvenile phase of orchards by producing quality planting materials.
- It is only authenticated sources for getting best planting materials.
- Controlled transit of planting materials.

Demerits

- License validity is for short period.
- Getting license is very lengthy and time consuming procedure.
- Inspection fees of the Government officials are very high.
- Poor or inadequate inspections by the agencies.
- Lack of test of genuinity of planting materials.
- Lack of skilled labour.
- Lack of knowledge among farmers and growers regarding planting materials.

Fruit And Ornamental Plant Nurseries (Regulation) Act

Application Form II

License for establishing/conducting a fruit plant nursery under the Fruit and Ornamental Plant Nurseries (Regulation) Act, 1995.

License No. ____________________ Date of issue ____________________

Son of ________________________ of village ______________________

Postoffice________________________Tehsil________________________

District _____________________Owner of ________________________

is hereby authorized to raise, exhibit for sale and self for transplantation of fruit plants of the following kinds and varieties:

1.

2.

3.

4.

5.

6.

7.

8.

This license is valid from ________________ to______________________

The license shall be subject to the following conditions:

1. Specifically, the licensee must comply with section 8 (reproduced below) of the Act and shall not contravene any of the provisions of the Act or the rules made thereunder.
2. Licensees must conduct their business honestly and fairly.
3. On demand of the competent authority or any person authorized by it, the licensee shall produce his license, the registers and other records required to be kept under this Act and the rules there under.
4. The licensee shall not permit any evasion or infringement of the Act or its regulations, and shall report in writing to the competent authority any evasion or infringement he becomes aware of.

5. Upon receiving instructions from the competent authority or any person authorized by it, the licensee shall promptly follow them.
6. When a licensee transfers control of the fruit Nursery in whole or in part, he must notify the competent authority within one month after the transfer.

Signature of Competent authority
With seal of his office

Period of: From ______________ to ______________Renewal ___________

Signature of competent authority
With seal of his office

Nursery accreditation

Fill out the online application form and attach the required documents (refer checklist down load from www.nhb.gov.in) and submit the application to National Horticulture Board for accreditation (Recognition and Renewal) in detail as follows:

- A layout of the nursery showing the locations of the infrastructure components
- Plan for land utilization
- Details of qualification of technical staff
- Major farm machinery equipment
- Prepared a manual for operation
- Planting material production process
- Planting material inventory management
- Details regarding source of mother plants used for propagation
- Register for sale of horticulture plants

Assessment criteria: System of graded certification

1. Different from licensing of horticulture nurseries under provision of some Act or administrative orders.

2. It is based on
    - Sources of parent material are continuously evaluated
    - Plant propagation in disease-free conditions
    - Good nursery management practices
    - Maintaining reliable records
    - Staff training
3. Each parameter will be critically examined by the assessment team according to the criteria laid out

References

Bhardwaj, R. L. and Sarolia, D. K. 2011. Modern Nursery Management. Agrobios (INDIA). pp. 14-27.

Bhujbal, B. 2011. Resource Book on Horticulture Nursery Management. The Registrar, Yashwantrao Chavan Maharashtra Open University, Nashik, pp.42 -52.

Davidson, H., Mecklenburg, R. and Peterson, C. 2000. Nursery Management: Administration and Culture. 4th edition, Prentice Hall, pp. 96-118.

http://ecoursesonline.iasri.res.in/mod/page/view.php?id=148506.

http://nhb.gov.in/nurindexpage.aspx?tab=41.

Sarolia, D. K., Samadia, D. K., Choudhary, B. R. and Singh, D. 2018. Production of Quality Seed and Planting Materials. CIAH/Tech./Pub. No.75. ICAR-Central Institute for Arid Horticulture, Bikaner, Rajasthan (Indian): 38

13

Import and Export of Seeds and Planting Material and Quarantine

Kapil Mohan Sharma, R.K. Jat and M.L. Jat

1. Determination of qualitative nursery planting materials

1.1 Quality assessment of seedlings

Nursery is a place where young/infant seedlings are kept under intensive care until they are ready for planting. The production of quality seedlings begins with the collection of quality seed, nursery establishment, and after-germination maintenance. The quality of plant propagation structures may not be a qualitatively gradable criterion, but it can be indirectly assessed by their relationship to growth, productivity, and vigour. The Dickson Quality Index (DQI) is very useful for assessing the quality of a specific group of seedlings. This is done by selecting some seedlings at random from that group and computing Dickson Quality Index (DQI).

$$DQI = \frac{\text{Total seedling dry weight (g)}}{\{\text{Height (cm) /stem diameter (mm)}\} + \{\text{shoot dry weight (g)/root dry weight (g)}\}}$$

Limitations of DQI

1. The quality of seedlings of even age can be evaluated; comparisons of seedlings of different ages are not available.
2. The DQI is derived after destructive sampling. Therefore, its implications for live seedlings may be appropriate only for academic purposes.

Bureau of Indian Standards (BIS-2008) related to nursery

Nursery quality measures can be controlled by different systems and standards. Due to the highly localized nature of agriculture, cultural practices and varietal preferences vary widely from region to region. In addition, the opening up of the world market has led to an increase in the trade of agricultural products. In order to facilitate trade in these products and gain the confidence of consumers within and outside the country, it is imperative to define and assign certain common minimum standards.

General quality standards for nursery plants

1. It is important for nursery plants to have an appropriate ratio between shoots and roots.
2. It is imperative that nursery plants are weed-free. In accordance with variety and species, leaf color and morphology should be standard.
3. Ideally, the nursery plant should be free of disease and pests and grow vigorously.
4. The graft union should be healthy, and the scion and rootstock should be equally sized.
5. During shifting and transportation, seedlings should not exhibit symptoms such as leaf drying, yellowing, stress, *etc.*

Criteria for quality planting material of some important plant species

Crop	Methods	Quality standard
Aonla	Budding	6-12 month old
Arecanut	Seed	15-18 month old
Banana	Sucker/tissue culture	2-3 month old
Ber	Budding	1 year old
Cashew nut	Epicotyls/softwood grafting	5-6 month old propagules
Citrus	Grafting	1-2 year old graft, 75-90 cm in height
Coconut	Seed	1-1.5 year old, stem girth 10-12 cm, leaf stalk thick and short
Custard apple	Seed	5-6 month old
	Grafting	1 year old graft
Fig	Cutting	8-12 month old, 4-6 buds, 30-40 cm height, 1-1.25 cm diameter of propagules
Grape	Cutting	15-20 long, 3-4 buds, 2.5 cm diameter of propagules
Guava	Layering	6-9 month old propagules
Papaya	Seed	15-22.5 cm tall, 1.5-2 month old
Pomegranate	Air layering	20-25 cm height, 3-6 month old
Sweet orange	Budding	150 days old
Sapota	Approach grafting	60 days after removal from the mother plant, leaves fully turn green

Import and export policy on seeds and planting material

The import and export of seed and planting material is governed by Export and Import Policy 2002-07. Based on this an EXIM committee has been formed under the Seeds Division to deal with seeds and planting material in accordance with the New Policy on Seed Development (1988) and Plant Quarantine Order (2003). The importer or exporter needs to import the EXIM committee in advance about the trade in prescribed format and the committee review the application and approve the transaction. The import licenses are granted by the office of Directorate General of Foreign Trade (DGFT) after the consultation with Department of Agriculture and Cooperation (DAC). ICAR and State Agriculture University can import seeds/planting materials for trial and experiment purposes but for commercial licenses trial/evaluation report of the seed performance and their resistance to seed/soil borne diseases is required. For this the importers need to import a small sample of seeds for testing/accession at NBPGR. The imported seeds/ planting material need to either accepted or rejected by Plant Protection Advisor after quarantine checks within prescribed time limit. The duration of the quarantine period varies from crop to crop. The details of the policy and updated changes are available at https://seednet.gov.in.

According to the Wood Seed Trade Statistics, India is the sixth largest market for domestic seeds in the world which is estimated to be about 1300 million dollars. To boost India's seed export India has participated in The Organization for Economic Cooperation and Development (OECD), an international scheme which allows common certification rules for its members countries. The compilation of the technical regulations established by participants states are made available to smooth functioning of their national sectors and foreign trade. Currently, India has decided to participate in OECD schemes of following categories of crops

1. Grasses and legumes
2. Crucifers and other oil and fibre species
3. Cereals
4. Maize and sorghum
5. Vegetables

Role of plant quarantine service

Plant quarantine service was established to prevent the entry, establishment and spread pests in India as per the provisions of the Destructive Insects and Pests Act, 1914. It helps to improve productivity by preventing entry of foreign pests and evasive plants. Additionally, they facilitate international trade in agricultural products by facilitating export certification of plants and plant products. This helps in the adoption of safe quarantine practices to protect our environment.

New seed policy

"New Policy on Seed Development" was issued on September, 16, 1988 with the objective of supplying farmers with the best planting materials available in order to increase productivity, farm income, and export earnings. The policy covers coarse cereals/ pulses/oil seeds; vegetables and flower seeds; bulb/ tubers and flowers; cutting/sapling/bud/ wood *etc.* of flowers; and seeds and planting material of fruits.

According to the New Seed Policy, there are prescribed guidelines for import of seed/ planting materials.

1. Import of seeds of coarse cereals/pulses/oil seeds

- By collaborating with companies abroad, companies can import seeds for a period of not more than two years if they provide parental lines/ nuclei or breeder seeds.
- For ICAR multi-location trials, bulk import is permitted on recommendation of the DAC.
- The detailed testing of imported consignments shall be carried out for 30-35 days for the same purpose, and until then it will be held in the AAI warehouse or customs warehouse at the expense of the importer.

2. Import of vegetables and flower seeds

- Bulk import is permitted under Open General License (OGL) by eligible importers *viz.*

   i. Department of Agriculture/Horticulture of State Government, State Agricultural Universities and ICAR.

   ii. Seeds producing Indian companies/firms registered with National Seed Corporation.

   iii. National Seed Corporation, State Seeds Corporation .

   iv. Food Processing Industrial Units.

   v. Growers of vegetables and flowers registered with Director of Horticulture/Agriculture of State Government.

- The detailed seed testing of imported seeds will be done for 30-45 days on arrival of port of entry. Permit by PPA is required only for vegetable seeds and not for flower seeds.

3. Import of bulbs and tubers of flowers

- Bulk import of bulbs and tubers of flowers and ornamentals is allowed under OGL by eligible importers.
- Grow out test of the imported consignment will be done for 30-45 days on arrival of port of entry and the consignment will be held under detention in cold storage or may be kept under post entry quarantine (PEQ) instead of grow out test on request of importer.
- Bulbs/ tubers are grown in individual poly bags and are evaluated by DIA and PQ officers as per the specified time of permit.

4. Import of cuttings/saplings/bud wood *etc.* of flowers

- OGL allows the importation of flowers cut, saplings, bud wood, *etc.* However, permit is needed by PPA or competent authority.
- An importer must establish a PEQ facility before importing and obtain approval from a Designated Inspection Authority (DIA).
- Plants under PEQ are approved not exceeding 45 days.

5. Import of seeds/planting materials of fruits

- PPA allows the import of seeds/planting materials of fruit plant species on a case-by-case basis on recommendation of the state Director of Horticulture/ Agriculture and subject to quarantine regulations.
- Permit of PPA or competent officer is required for the import of the same.

Procedure for import and export of planting materials

Import procedure: This involves two phases

Phase I: Issue of import permit: An importer intending to import agricultural commodities has to apply in advance for the issue of import permit in respect of the commodities listed in Schedule V and VI of PQ Order, 2003 in the prescribed form. Afterwards, inspection of imported agricultural commodities on arrival at the port of entry for preventing the introduction of exotic pests and diseases inimical to Indian Fauna and Flora through implementation of DIP Act, 1914 and Plant Quarantine (Regulation of Import into India) Order, 2003 issued thereunder:

During the import clearance process, various steps are taken such as receiving a reference from customs, sampling, detailed testing *viz.*, bacteriological, mycological, entomological, nematological, *etc.*, besides the post entry quarantine (PEQ) testing at the importers premises under the PEQ facility. The post-entry quarantine inspections required for cuttings, saplings and bud wood are conducted by the Designated Inspection Authorities constituting mainly the head of the Department of Entomology/Plant Pathology of the State Agricultural Universities/ICAR Institutions.

Phase II: Undertaking post entry quarantine inspection in respect of identified planting materials: According to the New Policy on Seed Development, 1988 and PQ Order, 2003, specified propagation materials (*viz.*, cuttings, saplings, bud woods, *etc.*) require growing under Post Entry Quarantine for a specified period. Such planting material can be imported with a permit provided by a certificate from the Designated Inspection Authorities of the concerned jurisdiction stating that the importer possesses post-entry quarantine facilities for the imported planting material. Consignments of this type are released after informing the relevant Inspection Authorities for performing further inspections, and the final clearance is granted based on the inspection report.

Export inspection and certification procedure

Inspection of agricultural commodities meant for export are done as the requirements of importing countries under International Plant Protection Convention (IPPC) 1951 of UN-FAO. The export inspection includes sampling of detailed laboratory tests in case of seeds and planting materials for propagation whereas visual examination with hand lens and washing tests, *etc.* are carried out for plant material meant for consumption. The export inspections are conducted at exporter's premises also to facilitate exports for agricultural commodities meant for consumption. However, there are various steps involved for inspection and certification of plants and plant products.

1. Registration of application

- Exporter need to submit application in prescribed format to PQ station of designated port through which s/he intends to export.
- The importing authority must provide a copy of the permit.
- The PQ authority will examine the application and proceed with the necessary formalities.

2. Inspection/ sampling and laboratory testing

- It is the exporter's responsibility to show the consignment either at the office of a PQ station or arrange for inspection at his premises.
- In order to propagate seeds, they must be sampled according to the rules of the International Seed Testing Association (ISTA), 1976.

3. Fumigation and treatment of consignment

- In the event of live insect infestation is notified, the exporter shall arrange for fumigation of consignment or container at this premises as approved place by pest control operator.
- After fumigation, the consignment is re-inspected for live infestations.

4. Issuance/ Rejection of Phytosanitary Certificate (PSC)

- Phytosanitary certificate is issued if the consignment is found to be free of quarantine pests.
- The issue of PSC will be rejected if the commodity on inspection is a prohibited one, found to be affected by quarantine pests, or cannot be fumigated to make it pest-free.
- PSCs can also be rejected if the commodities contain objectionable plant materials or are contaminated with soil or noxious weed seeds.

References

Anonymous, 2008. Export/import policy on seed and planting materials. Department of Agriculture and Cooperation, Government of India. URL: http://www.seednet.gov.in/PDFFILES/Brief%20on%20Export_Import%20Policy%202002-07.pdf (Assessed on 08/04/2022).

Anonymous, 2015. Plant Quarantine (Regulation of Import into India) Order, 2003, and includes amendments issues thereto from time to time. Directorate of Plant Protection, Quarantine and Storage, Government of India. p.314.

Anonymous, 2020. Export Inspection and Certification. Directorate of Plant Protection, Quarantine and Storage, Government of India. URL: https://plantquarantineindia.nic.in/PQISPub/html/Exp-insp-cert.htm (Assessed on 08/04/2022).

Anonymous, 2020. Import and Export Procedure. Directorate of Plant Protection, Quarantine and Storage, Government of India. URL: http://ppqs.gov.in/divisions/plant-quarantine/import-export-procedure (Assessed on 08/04/2022).

Anonymous, 2022. OECD seed schemes: Rules and Regulations. OECD. URL: http://www.oecd.org/agriculture/code/seeds.htm (Assessed on 08/04/2022).

Ratha Krishnan, P., Kalia, Rajwant K., Tewari, J.C. and Roy, M.M. 2014. Plant Nursery Management: Principles and Practices. Central Arid Zone Research Institute, Jodhpur. p.40.

14

Marketing of Nursery Plants

R.K. Jat and M.L. Jat

The main purpose of establishing a nursery is to supply plants whether to one's own garden or to out station indentors to fetch good income. To obtain adequate returns on the investment made by a nurseryman, one should have fair idea about lifting, packing and storage of the nursery plants.

Lifting of nursery plants

The plants grown in the nursery are to be lifted for establishment at their permanent place or for sale purpose. The careless lifting may affect adversely the establishment of nursery plants. For getting higher rate of success of evergreen plants in the field, it is always advisable to lift the saplings with a ball of earth intact particularly for mango, guava, citrus and litchi *etc.* However, the deciduous plants like peach, grape, apple, kiwi fruit, hazelnut, pecan nut, cherry, almond, walnut *etc.* can be lifted bare rooted during dormant conditions. If possible, it is advisable to irrigate the nursery 3-4 days before the actual date of removing the plants in order to prevent root damage. The lifting of evergreen saplings during wet weather is not recommended because earth balls are not formed properly as a result of excessive soil moisture or heavy rain.

The uprooted nursery plants should be inspected by a team of experts for parameters such as insect attack, disease infestation, healthy root system, and other pomological characteristics. Before packing and storing seedlings, unhealthy saplings should be discarded.

For proper identification, the saplings should be labelled crop-by-crop and variety-by-variety with zinc labels, plastic labels, or wooden labels, and labelled for nursery plant sale.

Packing of nursery plants

The seedlings/saplings should be properly packed after lifting and labelling. Packing is the process or method of keeping the young plants together until they can be transplanted. So, they must be packed so that they do not lose their turgidity and are able to establish themselves on the new site.The type of packing depends on the size of stock and type of plant. The bare rooted saplings should be packed in moist sphagnum moss grass and wrapped with hessian cloth to avoid drying of the roots. Before placing the saplings of evergreen plants into polyethylene bags and then keeping them in baskets, they should be wrapped in coconut leaves or paddy straw. For long distance transport, strawberry runners and vegetable seedlings are first packed in moistened moss grass and stored in hessian cloth bags or in baskets.

Generally, considerable focus should be given while nurseryman sends plants to a distant place but it does not mean that the plants to be sent to nearby places should not be packed properly. Nurseryman should pay more attention to the packaging of the material to ensure a better price. A packing material should have the following characteristics:

- It must be readily available, cheap, and perfectly suited to the material being packed.
- It must look attractive to the indentors.
- It must not affect the quality of plants to be delivered conveniently.
- It must ensure safety against desiccating and mechanical injury.
- It should be well adapted to convenient and economical handling.
- It should be suitable for transport, loading with security and efficiency in a small space.
- It should not make the material bulky rather it should add minimum weight to the material.

Packing of nursery planting materials should be completed carefully so that their supply may be done in safe and sound condition. Nurseryman must avoid bruising when placing planting material into a package. The joints of grafted and budded plants should be taken care of to avoid breaking. For distant places, most of planting materials is sent by truck and rail, where it is handled very roughly. Therefore, proper packing of planting materials is more relevant under such conditions.

Materials used for packing

Nursery plants are packed in different ways using the following types of materials:

Hessian cloth	:	Made from the good quality jute fibers
Sacking cloth	:	Made from the raw grade jute fibers
Plastics	:	Low & high density polyethylene, polypropylene, nylon
Bottles or Tins	:	For storage of clean seeds
Sphagnum moss	:	For wrapping the earthen ball of the saplings
Dried grass	:	For wrapping the earthen ball of the saplings
Moistened moss grass	:	For wrapping up the delicate planting material *viz.,* strawberry runners and vegetable seedlings before packing
Paddy & wheat straw as well as banana & coconut leaves	:	For wrapping the earthen ball of the saplings

Storage of saplings

The high or low temperature and humidity may favour desiccation of the plants lifted from the nursery. Planting materials should be kept in a storehouse before sending them to distant places. The storehouse should have proper ventilation which helps in driving out the carbon dioxide and heat of respiration. Low temperature (0.5 to 3.0°C) in association with high relative humidity (85-90 %) in the atmosphere of storage reduces the respiration and transpiration rates.

Transportation of nursery plants

Adequate transport facilities are of prime importance for successful nursery business. A nursery should be located near a metal road, or close to a railway station to deliver the plant materials to the customers in a short time. It should be easily accessible to the customers, but totally free from pilferage. The nursery plants must be picked up the day and are received by the transport agency. There are no proper seedling storage facilities at the transport agencies, so seedlings deteriorate rapidly in these conditions. Planting materials should be delivered from transport and planted in the permanent field or place as quickly as possible, ideally within 24-72 hours of delivery. An enclosed vehicle must be used to transport planting materials. If the vehicle is open back it is covered with a tarpaulin to protect the planting materials from direct sun and drying in the wind. Some ideas should be kept in the mind during the transportation. The vehicle should never be parked in the direct sun during transportation of planting materials. Planting materials packed inside of boxes or bales can reach harmful temperatures when exposed to the sun. The planting materials may be damaged from brushing when these boxes are to be dropped.

Care of plants

After receiving the plants from transportation, they should be stored in a shady, calm environment so that the heat absorbed from the transportation can be lost. It is important to confirm the variety, number of plants packed, *etc*. on the label. Sprinkling water on the leaves and roots will keep them moist. The plants should not be stored at field condition.

Sale of planting material

Marketing of nursery products has been a great problem in India because there is no appropriate marketing system in our country. A market includes sales and purchases. Before starting a nursery, it is important to ensure that there is an outlet for the produce in the area. A marketing strategy at a distance will swallow up profit because of the high transportation cost involved, along with damage to the plants and other losses.

Nursery plants are generally in high demand during rainy season. In order to do so, advertisements in local daily newspapers, posters, handbills, catalogs, and commission agents can all be used.

Precautions

- Plant roots should not be damaged while uprooting nursery plants.
- To ensure that plants offered for sale are true to type, they should be uniform.
- Nursery plants were inspected without negligence.
- Nursery plants should be packed carefully to prevent root system drying.

References

Bhujbal, B. 2011. Resource Book on Horticulture Nursery Management. The Registrar, Yashwantrao Chavan Maharashtra Open University, Nashik, pp.107 -109.

http://ecoursesonline.iasri.res.in/mod/page/view.php?id=96886

Sharma, R. R. 2002. Propagation of Horticultural Crops: Principles and Practices. Kalyani Publishers, New Delhi.

15

Tools and Accessories for Nursery

Vivek Saurabh and Ashok Dhakad

Introduction

The nursery is an area where tender seedlings are raised and conditioned before planting out in transplant beds or directly in the field. The increasing demand for quality planting material has up stepped the nursery business in many countries. The sophisticated technology, research and novel production approaches has transformed the traditional horticultural nursery to a technology intensive business. The tools and accessories associated with nursery production helps in aiding productivity of the nursery. The colour-coded and labelled plant materials, textured pots *etc.* are fascinating to consumers, indirectly increasing the salability of the plants and profitability of the enterprise. The well-established nurseries nowadays rely more on automated technologies like potting machines, tray fillers, automated humidity control system *etc.* Use of tools and equipment helps to reduce the drudgery for the employee without compromising with their safety. It improves efficiency of different nursery management operations. This chapter throws light on the existing equipment and accessories associated with nursery management.

1. Types of nursery equipment

The tools and implements have played a pivotal role in the fuelling the agricultural revolution. The pre historic hoe, a mere V-shaped iron piece has paved way to hundreds of other tools we witness around.

1.1 Tools

Tools are mostly operated manually. However, semi-automatic versions do exist. Design of the tools, material, strength and sturdiness of the tool are few pre-requisites that should be kept in the mind before inducting them into nursery.

Before tool selection it is important to keep in mind:

- **Design** – This should be comfortable, if the handle is wooden it makes sure it is free of splinters therefore offering proper grip.
- **Durability** – Tools should be able to withstand heavy wear and tear that arises as a result of regular operations.
- **Quality** – For heavier garden operations the quality of tool holds a prime importance. Tools made of stainless steel or aluminum is corrosion free. Similarly, well-designed tools tend to cost more but are user friendly. The nursery if set up in area with clayey soil, requires good quality, heavy duty tools.

1.2 Types

1.2.1 Secateurs

Secateurs are modified version of scissors, primarily used for cutting or pruning plant stems up to 1–1.5 cm (0.5"). It usually contains blade made up of tempered carbon steel and a sturdy aluminum alloy handle coated with vinyl. On the basis of cutting edge secateurs are categorized into two types-

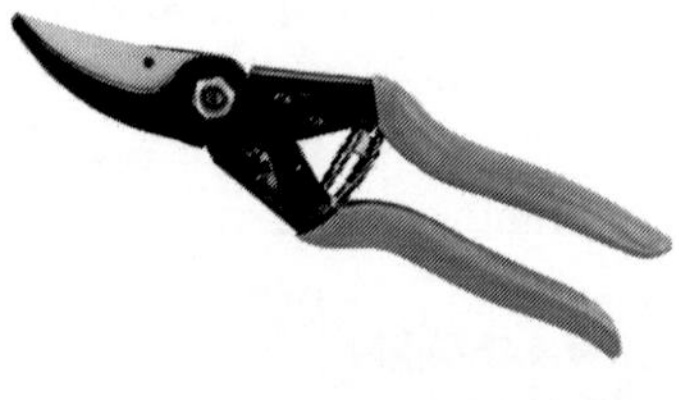

a) **Scissor type** – It gives a clean sharp cut.

b) **Anvil type** – It rather has chopping action and tends to bruise the plant as the anvil prevents a close cut. The cut is not clear. These are usually used for removing tough dead wood.

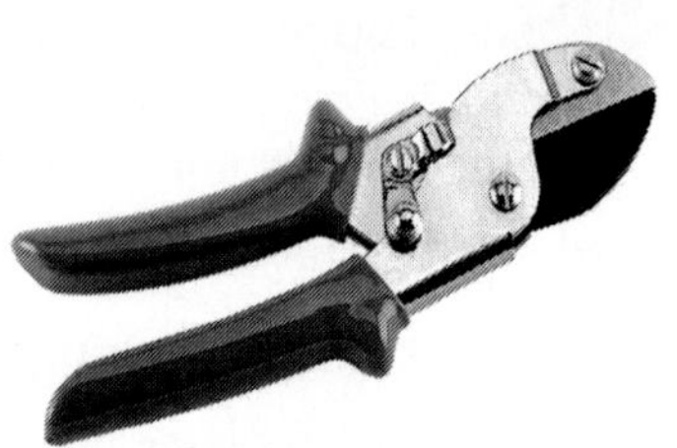

Secateurs

Handling

- Sharp and clean cut, preferably above the node should be made. It prevents bruising of stem and prevents infection.
- Sharpen the blade whenever necessary and keep it clean and oiled after every use.
- Not to be used for pruning or thinning out thick stems as it will strain the blade.

1.2.2 Knives

Customarily, knives are used for pruning branches, budding, grafting, dividing plant material *etc.* Furthermore, knives can have a fixed or a movable/ folding blade. Nurserymen often use a folding blade knife. The nursery knives can be of different types depending upon the purpose they are being used for. Weight of the knife should be balanced on finger to make sure the weight is towards the handle. A comfortable handle is a must.

a) **Pruning knife** – It is used for pruning the small shoots. Fixed blade knife is usually used for pruning purpose to prevent any accident associated with mishandling.

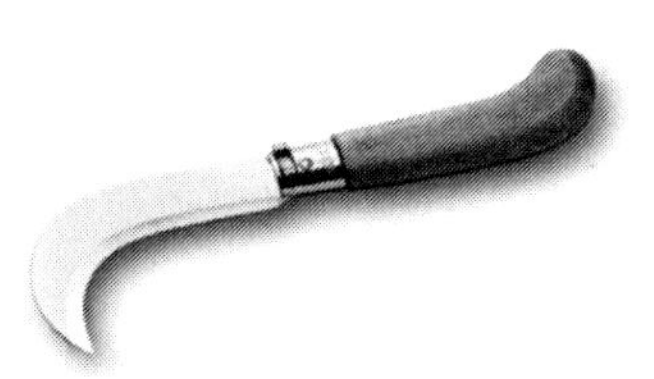

Pruning knife

b) **Budding knife** – These knives have a comparatively shorter curved apex which facilitates the easy incision on the root stock. The rear end of knife comes with a spatula shaped blade for opening the bark and the placement of the bud. Budding knives with double blades are often used for patch budding.

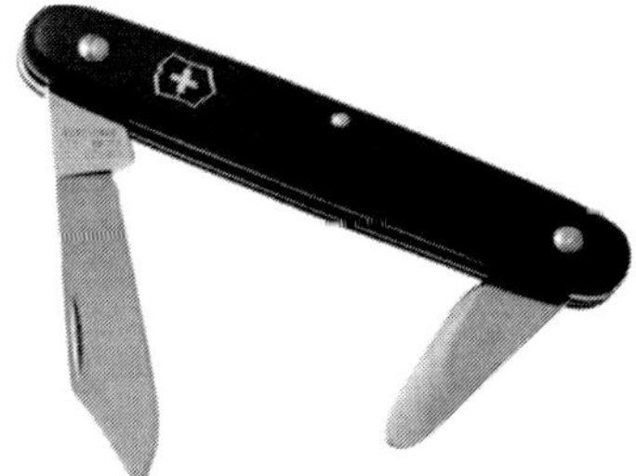

Budding knife

c) **Grafting knife** – It is made up of long straight blade of almost 7-8 cm and a strong handle. It aids detachment of scion from mother plant, its defoliation and making incision on the root stock.

Grafting knife

1.2.3 Saws

Saws have long narrow blade to make way through the narrow crotch angles of branches, with widely spaced serrated edge. These teeth facilitate lopping of green wood. Different variants of the blades are available. Half moon saws have circular blade, Bow saws are bow shaped and have well-spaced teeth and carpenter's saw are straight edged with finer teeth to give a clean narrow cut. However, Bow saws are the most popular as light pruning saws.

Handling

- The cutting edge should be sharpened from time to time for unimpeded and clean cutting experience.
- Saws should be stored vertically as lying against the floor may affect the quality of the blade.
- Blades suffer great deal of damage if rubbed against gravel or dropped on floor.

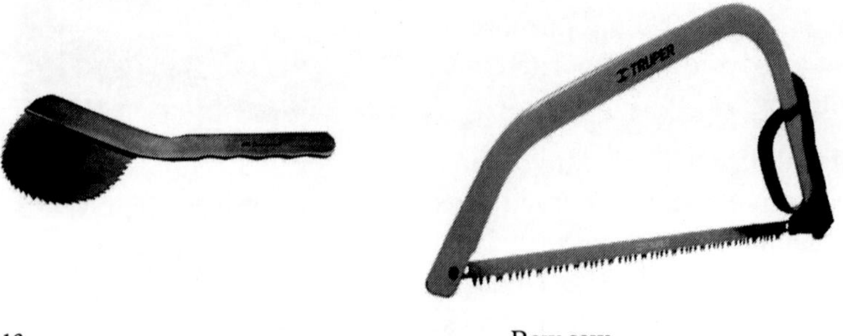

Half moon saw

Bow saw

Carpenter's saw

1.2.4 Spades

Spades, in general are used for digging and planting. It primarily comes in two variants- a short handled one, which is mainly employed for digging around garden beds and another with a long handle, which is most commonly used for digging deep trenches and constructing irrigation channel. It can also be used for lifting and turning of soil.

Spade

1.2.5 Shovels

Shovels are not shaped to dig; the main purpose of the tool is to transfer the dug out soil within the field for leveling or other on farm operations. An inwardly curved spoon-shaped blade protrudes from the handle. A shovel is not designed to dig.

Shovel

1.2.6 Forks

This tool is designed for on ground operations like – breaking up of compact soil, digging out the rocky soil and turning of lighter soil. It is a four-pronged tool usually constructed with high carbon steel and has a long handle attached for easy working. The forks can be further categorized into:

a) **Garden forks-** It is usually employed for heavy gardening operation in spacious nursery/ garden. Aeration, breaking hard clods of soil, double digging *etc.* are some of those operations.

b) **Digging fork-** These are also four tined forks used for comparatively lighter manoeuvre like digging or harvesting tubers *etc.*

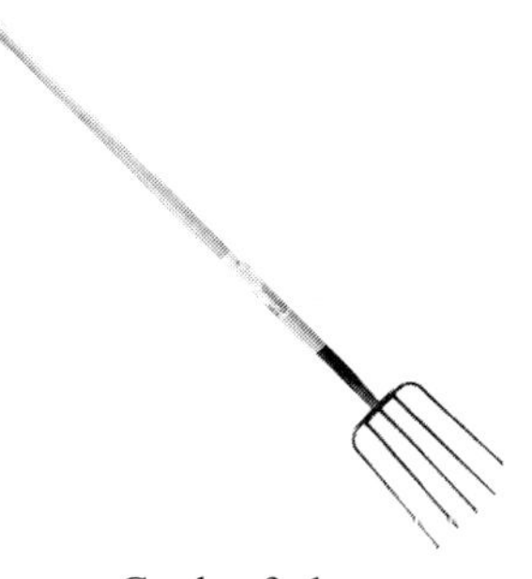

Garden fork

c) **Border forks-** It only differs in size from the above mentioned forks and is used in gardens where space is a constraint.

1.2.7 Hoes

Hoes are utilized for weeding out the rows of standing crops without damaging the crop stand. It further enables pulverization of soil and loosens it thereby aiding aeration. The common type's available are-

a) **Straight blade hand hoe-** It has a bar shaped, wider downward facing blade to remove weeds from inter row space.

b) **V-blade hand hoe**- It has a long handle and a curved V shaped blade. It is operated while in standing position. The pulling action leads to uprooting of weeds and also creates shallow furrows.

c) **Dutch hoe-** It has a shallow angled blade with a front cutting edge.

d) **Karjat hoe**- It has three V shaped blades assembled together in staggered pattern and is operated same as V blade hand hoe.

e) **Three tined hand hoe-** It is a long handled, 3 pronged hand hoe employed for weeding in line sown crops.

1.2.8 Rakes

Rakes possess long flat teeth like projections, which create a brushing effect over the surface. However, the shape and size of the tines can be modified depending upon the purpose. The metal rakes are mostly preferable owing to its durability. It is a multi-utility nursery tool that serves the hoeing, earthing, leveling and collecting loose grass clippings or weeds. The adjustable rakes available in markets can be modified according to size of the material being raked.

Rake

1.2.9 Scissors

It is shearing tool that consists of pair of edged metal blades pivoted at a point such that blades slide together. Scissors can be used for removing flower parts or cutting flowers and other unwanted tender plant parts.

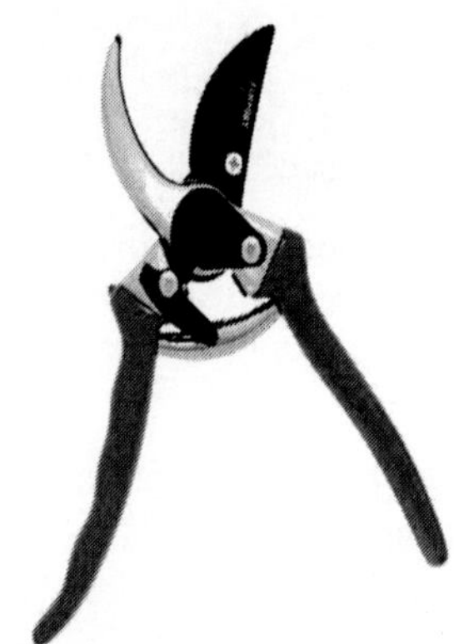

Scissor

1.2.10 Wheel barrow

Wheel barrow serves as a load carrier in gardening/ nursery operations. It usually contains a single wheel towards the center of the cart and two handles at the rear end which are supported by two legs. It helps to carry nursery plants, fertilizers *etc.* from one place to another on farms.

Handling and care

- Center of mass of the barrow should always lie over the wheels and not on the handles.
- While wheeling, ample leg space should be available to the user.
- The barrow should have strong frame. A well-greased barrow serves longer.

Wheel barrow

1.2.11 Automated equipment/ machines

1.2.11.1 Seeding machines cum tray filling machines

Nursery seeding machines are mostly used for developing nursery transplants of vegetables like tomato, cucumber or flowers. Soil is filled up automatically into the trays and seed is dropped utilizing vacuum mechanism aided by sensors for precision. An indigenous, low cost tray based nursery seeding machine was developed by (Jadhav *et al.,* 2020) MMCE Pune India.

These machines help in cutting down the labour charges and prove cost effective in long run. It promotes judicious use of resources by minimizing the wastage thereby increasing the productivity.

1.2.11.2 Potting machines

Potting machines are usually based on an electro-pneumatic mechanism for filling up the pots. It can be manual, semi-automatic or automatic depending upon the need of the user.

Machine consists of a conveyor, media storage area, volume control set up *etc.* The potting rates vary between 200- 500 per hour work.

Potting machine

1.2.11.3 Heating trays

These trays have heating cable made up of metal of high resistivity either sandwiched between two metal layers or placed on wire mesh. Heating mats are also available that provide suitable temperature for germination of seed. The lid of the tray provides a safe environment protecting the germinating seeds from insects or pests. Existing designs can be further modified to suit the nursery needs.

1.2.11.4 Mulchers

Mulchers are helpful in shredding the over grown vegetation and pruning refuse. It can be attached to a rotator mower or a tractor or power operated leaf mulcher or shredder for small scale gardens. Shape of blade varies from disc to tooth shaped. However, for chipping the woody refuse a relatively powerful chipper needs to be installed. These shreds and chippings can be mixed with potting media. Furthermore, it can be used as organic mulch for nursery beds and can be added to compost.

Mulcher

1.2.11.5 Handling

- All the moving parts should be well-greased.
- Only plant material should be fed into the hopper.
- Ear-muffles can be worn by the operator.
- Must be operated by an adult.

2. Accessories: These are commonly in usage.

2.1 Containers/ pots

Pots or containers are an important accessory as it is used for propagation, display, growth and transport of garden plant. Containers made up of plastic, terracotta, fiberglass, metals or bio materials like coir *etc.* are available in market. They occur in variety of shape, size and texture. The pots or containers must have certain characteristics which are as follows:-

- Strong and durable material.
- Design that minimizes ringing or matting of roots.
- Space efficient.
- Proper drainage holes to prevent wetness of soil and promote proper air to water ratio.

Table 1: Types of container materials

S.No.	Type of material	Characteristics	Disadvantages
1	Plastic	Manufactured in variety of colours and textures; retains moisture for longer period; economical.	Discolored under prolonged exposure to sunlight; may develop crack over time; Non-biodegradable.
2	Terracotta	Durable, sturdy and long- lasting; porous nature allows efficient air and water circulation; good for succulents and cacti.	Requires frequent watering; unglazed pots may spill water from the sides; costly when compared to plastic containers.
3	Fibreglass	High durability, easy to clean	Medium life; may discolour over time.
4	Concrete	Durable; insulates plant against sudden temperature fluctuations; can be decorated in several ways.	Leaching of lime from concrete; soil dries quickly due to porous nature; heavy.
5	Brass	Eye catching; sturdy; suitable for indoor plants.	Needs polishing after a certain time period; expensive.
6	Timber	Natural and eco-friendly appeal	May rot over time making plant prone to diseases and pest.
7	Paper	Recyclable; Eco-friendly; suitable for nursery seedlings.	Limited life; cannot withstand heavy watering.

Pictures of various containers

Plastic pots

Terracota pots

Fiberglass pot

Concrete pots

Timber pot

Paper pot

2.2 Basket liners

Basket liners are particularly used to line the containers that may be open wire baskets or a closed container made up of plastic. It prevents the potting mixtures from draining along with water. The most commonly used basket liner is coir followed by sphagnum moss and fabric. Basket liner made up of Kenaf fiber has been found superior to coir and sphagnum moss as it keeps soil well hydrated and fully saturates it with water before draining excess water (Henson, 2005)

2.3 Labels

Labels act as quick guide to plant identification in a nursery or garden. Labels can be personalized and designed according to need of the nursery/ garden owner. Labels serve several purposes like-

- Categorizing and arranging nursery plants according to their botany, nature and growth habit *etc.*
- Helps lay-man to identify different plants.
- Helps in labelling quality plant material.
- Maintains the nursery in order and keeps track of planting material available at nursery.
- Imparts tinge of professionalism to the nursery/ garden.
- Aids proper scrutiny of the plants.

In recent past, labels have emerged as an important marketing tool. Vibrant and informative labels assorted into several shapes and sizes help in promoting nursery and improve the salability of plant material. The labels can be printed or non- printed blank labels depending upon the necessity; it should be resistant to wear and tear mainly due to harsh weather. Different kinds of labels we may come across in market are-

2.3.1 Customized labels

These are high quality labels customized for promoting specific plants. Holograms or foils can be used.

2.3.2 Descriptive labels

It consists of details like accession number, common name, family *etc.* or may contains details like- sunshine requirement, care and maintenance, planting process *etc.*

2.3.3 Slip on tags

The ends of label slip on each other and get fastened. The information is printed with a thermo-stable ink that doesn't fade despite of odd weather conditions.

2.3.4 T-labels

It has a rectangular/ square shape for labelling and the pointed end is pushed into soil. T-Labels can be comfortably used for seed trays and pot plants.

2.3.5 Adhesive labels

These are coated labels with potent adhesives. These are mainly used for labelling pots and seeding trays. The coating prevents it from water and light in long run.

2.4 Staking accessories

2.4.1 Trellis and obelisks

Trellis is an architectural framework to support and display weak stemmed, tender climbing plants. It can be metallic, wooden or plastic in construction. Garden obelisks serve the same purpose but are tall frames with rectangular or square cross- section and a pyramidal top.

2.4.2 Stakes

Stakes are simply a wooden or metallic rod with peg like end which is fixed into the soil and individual plant is tied to the stake. It prevents the weak stemmed plant from drooping. It also provides support against gusts of wind. The staking accessory if to be installed for long period of time, should be durable and strong. However, for temporary stakes economical material can be preferred.

Main functions of staking accessories are as follows-

- To direct the plant into a particular shape.
- To contain the growth of haphazard bushes into an orderly trained framework.
- To support frail stemmed and droopy plant during early stage of its life.
- To alter growing habit of a plant.

References

Henson, D. 2005. U.S. Patent Application No. 10/922,552.

Jadhav, V., Kulkarni, R., Kotalwar, R., and Kunte, A. 2020. Automatic Nursery Seed Sowing Machine.

Mason, J. (Ed.). 2004. Nursery management. Landlinks Press.

Whitcomb, C. 2009. U.S. Patent No. 7,481,025. Washington, DC: U.S. Patent and Trademark Office.

16

Preparation and Management of Nursery Records

R.K. Jat and M.L. Jat

Introduction

The nurseryman needs some records if he is going to plan for remunerative returns and establish himself as a reputable business. Ideally, a nursery planner should be able to estimate how long it takes to lift saplings or how much seed is required, or how many cuttings he can accommodate in a glasshouse, as well as how much scion wood is required for grafting a particular plot. In nurseries, specific registers must be maintained to keep track of day-to-day activities, such as pedigrees and operations. In case of large nurseries, these records are required by law to satisfy inspection authorities, calculate taxes, detect theft, or discourage it.

Maintaining records is necessary for the following reasons:

- **Progeny orchard record:** A record, which shows the date of planting, source of mother trees, and layout of the different varieties of fruits, depicts the exact location of the progeny orchard. The variety additions and deletions must be documented chronologically. The quantity of scion wood taken from each mother plant should also be documented as well as the number of cuttings per plant and the total strength of mother plants needed to satisfy the demands of particular nurseries or nurseries generally.
- **Nursery layout record:** A plan detailing the layout of nursery plots according to crop type and within crop variety, mentioning the direction each plot begins or ends. Labelling the plots with tags or field plates is also recommended, if there is no risk of theft. As a result, a proper

record will be maintained along with the date of budding/grafting, days to sprout *etc.*

- **Cultural operations record:** The records of different agricultural operations including seed stratification, dates of sowing, germination, and farm operations such as dates of spraying, budding, grafting, fertilization dose and application. Other foliar sprays, including insecticides, fungicides, growth regulators, and nutrients *etc.*
- **Disposal register:** This register records all salable plants left over after thorough inspection. The records are kept crop-by-crop, variety-by-variety, in order to determine the anticipated output from the nursery and to assist with accounting.
- **Advisory/ visitor register:** In this register, experts should record their suggestions regarding a particular issue with their own hand writing. It is important to maintain a record of visitors with specific notes and feedback from the site. Experts and experienced nurserymen can contribute some valuable remarks.

The nursery entrepreneur should maintain records of sales and production of nursery plants, as well as records of expenditures in the nursery, which is called book keeping. Book keeping is one of the functions of financial accounting. It involves keeping records and books to record all the details of transactions made during the course of doing business. A business transaction can be categorized into several major activities/groups *e.g.,* sales, purchases, assets, *etc.* Separate books for recording transactions pertaining to these activities are maintained. Details of the transactions are recorded in the appropriate books. That process is known as book keeping.

Accounting records are important for the following reasons:

- Providing information and guidelines for nursery business planning, they provide up-to-date nursery business information.
- Their purpose is to analyze the performance of the nursery activity.

Various details of expenditures and income can be found in the following book records:

1. Purchase book

All transactions regarding purchases on credit or cash are recorded in this book. Purchase returns are also recorded separately in this book.

Date	Party"s Name	Bill No.	Ledger Folio	Item Name	Quantity	Rate	Amount	Terms
	Total							

2. Sales book

All credit or cash sales are recorded in this book. Transactions of sales returned are also recorded separately.

Date	Party"s Name	Bill No.	Ledger Folio	Item Name	Quantity	Rate	Amount	Terms
	Total							

3. Ledger

In this book, all accounts related to the transactions recorded in the journal or its subsidiary books are maintained, and postings are made as needed.

Debit Side Name of Account Credit Side							
Date	Particular	Folio No.	Amount	Date	Particular	Folio No.	Amount

It may be noticed from the format that a ledger account has two sides: a debit side (left hand side) and a credit side (right-hand side). There are also four sub-sections on each side, namely Date, Particulars, Journal Folio Number, and Amount.

(1) Date: In this column, the date of a transaction as it appears in the journal book from which the entry is entered into the ledger account.

(2) Particulars: This column lists the account in which the credit or debit of the subsequent transaction (under the double entry principle) occurs.

(3) Journal Folio Number: In this column the page number of the journal book or subsidiary book from where the transaction is brought to the account is mentioned.

(4) Amount: In this column the amount, with credited or debited from the account, appears.

4. Cash book

The cash book is the subsidiary ledger that maintains the account of 'cash'. It is also where transactions involving 'petty cash' are recorded. The cash book is nothing more than a cash account. Similarly to other asset accounts, this account must also appear in the ledger. Cash account, however, is not maintained in the general ledger, but is kept as a separate account and called cash book due to the multiplicity of cash transactions.

Debit Side (Payments)		Name of Account					Credit Side (Receipts)
Date	Particular	Journal Folio No.	Amount	Date	Particular	Journal Folio No.	Amount
					Closing Balance		
	Total				Total		

5. Bank book

The bank book is the subsidiary book of the ledger that keeps track of the account of the bank. A bank book is nothing more than a ledger that contains a bank account. Since the number of transactions involving banks is on the rise, it is convenient and appropriate to maintain a separate bank account where all transactions involving banks are recorded. Such an account is maintained separately and called a bank book. For this account as well as for other ledger accounts, all posting rules apply.

Debit Side Withdrawals)		Name of Account					Credit Side (Depositions)
Date	Particular	Journal Folio No.	Amount	Date	Particular	Journal Folio No.	Amount
					Closing Balance		
	Total				Total		

6. Stock register

A register is used to keep track of stock movements.

Date	Particular	Sales Book/ Purchase Book Folio No.	Receipts		Issues		Balance	
			Quantity	Value	Quantity	Value	Quantity	Value

The stock register is very similar to stock account. This shows us the closing stock available at the business to help the owner place further orders.

The registers and books should be issued through the Inventory Register so that the number of registers and books maintained can be tracked.

References

Bhujbal, B. 2011. Resource Book on Horticulture Nursery Management. The Registrar, Yashwantrao Chavan Maharashtra Open University, Nashik, pp.249 -252.

http://ecoursesonline.iasri.res.in/mod/page/view.php?id=96883